M000187199

Values at Work

Daniel C. Esty · Todd Cort
Editors

Values at Work

Sustainable Investing
and ESG Reporting

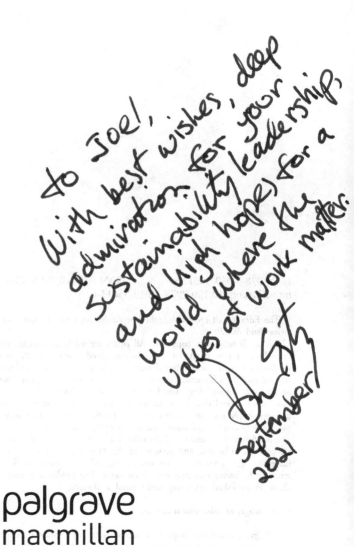

To Joel,
With best wishes, deep
admiration for your
sustainability leadership,
and high hopes for a
world where the
values at work matter.

Dan Esty
September
2021

palgrave
macmillan

Editors
Daniel C. Esty
Yale University
New Haven, CT, USA

Todd Cort
Yale University
New Haven, CT, USA

ISBN 978-3-030-55612-9   ISBN 978-3-030-55613-6   (eBook)
https://doi.org/10.1007/978-3-030-55613-6

© The Editor(s) (if applicable) and The Author(s), under exclusive license to Springer Nature
Switzerland AG 2020
This work is subject to copyright. All rights are solely and exclusively licensed by the Publisher, whether
the whole or part of the material is concerned, specifically the rights of translation, reprinting, reuse
of illustrations, recitation, broadcasting, reproduction on microfilms or in any other physical way, and
transmission or information storage and retrieval, electronic adaptation, computer software, or by similar
or dissimilar methodology now known or hereafter developed.
The use of general descriptive names, registered names, trademarks, service marks, etc. in this publication
does not imply, even in the absence of a specific statement, that such names are exempt from the relevant
protective laws and regulations and therefore free for general use.
The publisher, the authors and the editors are safe to assume that the advice and information in this book
are believed to be true and accurate at the date of publication. Neither the publisher nor the authors or
the editors give a warranty, expressed or implied, with respect to the material contained herein or for any
errors or omissions that may have been made. The publisher remains neutral with regard to jurisdictional
claims in published maps and institutional affiliations.

Cover image: © kaband\shutterstock.com

This Palgrave Macmillan imprint is published by the registered company Springer Nature Switzerland AG
The registered company address is: Gewerbestrasse 11, 6330 Cham, Switzerland

Values at Work *is dedicated to all those striving to harness the power of markets and finance to protect the natural world, promote human well-being, and create a more equitable economy and society.*

Acknowledgments

We are deeply grateful to a number of people and organizations that made *Values at Work* possible. First, of course, we would like to thank the 21 authors who contributed their time, expertise, and perspectives to this book. Their efforts within and insights about the field of sustainable finance demonstrate the leadership needed to create and fund a more sustainable planet. We are especially appreciative to our *Values at Work* editorial team led initially by Timothy J. Mason, Associate Director of the Yale Center for Environmental Law and Policy, and later by Tyler Yeargain and Maximilian W. Schreck, our Editorial Manager. Despite the COVID-19 pandemic, our group of Yale student editors excelled under pressure, ensuring the quality of the content in this book: Taryn Akiyama, Liam Chun Hong Gunn, Kara Hoving, Brett Cozzolino, Andy Xie, Omar Motala, Alexandra L. Wisner, Charles W. Harper, Robin Happel, Gillian Cowley, Maliya Ellis, Daiana Lilo, Ben Santhouse-James, and Margaret K. O'Leary.

Any project of this kind requires resources. AllianceBernstein, Barclays, Calvert, and Richmond Global Compass all provided critical sponsorship for the 2019 Yale Initiative on Sustainable Finance Annual Symposium, which served as the jumping off point for this volume. The Editors also received important support from the Yale School of the Environment, the Yale School of Management, and the Yale Law School as well as the Yale Center for Business and Environment and the Yale Center for Environmental Law and Policy. Additional resources were provided by supporters of the Yale Initiative on Sustainable Finance including Roberta Gordon and the Ida and Robert Gordon Family Foundation, NN Investment Partners, Vanguard, Mirova, and ClimateWorks Foundation.

We also benefitted from careful review and comments on the draft materials that became the chapters of this book from many colleagues and peers. In particular, we would like to thank the Advisory Board members of the Yale Center for Environmental Law and Policy, the Yale Center for Business and the Environment, and the Yale Initiative on Sustainable Finance for their feedback on the book's themes and content as well as their broader advice and support.

Finally, we want to express appreciation for all those who have been part of the Yale Initiative on Sustainable Finance programs and conversations on how to advance environmental, social, and governance (ESG) metrics and reporting as well as to understand the prospects for and obstacles to a world of expanded sustainable finance that would provide greater flows of capital to clean energy infrastructure and other elements of a sustainable future.

New Haven, CT   Daniel C. Esty
July 2020   Todd Cort

Further Praise for *Values at Work*

"With authors whose knowledge of sustainable investing and business practices is second to none, *Values at Work* cuts through the confusion and complexity surrounding ESG reporting to pinpoint what really matters, where the industry is going, and how serious investors can take full advantage of the trends."

—Martin Whittaker, *Chief Executive Officer, JUST Capital*

"The trend towards sustainable investing has accelerated. But measurable impact on the planet hasn't followed. *Values at Work* offers a clear call to action in outlining what's needed for the sustainable investment community to create meaningful change in the coming critical decade."

—Charlotte Kaiser, *Managing Director of NatureVest, The Nature Conservancy*

"At the heart of sustainable finance is the need to connect capital flows to the corporate actors who are true innovators in developing best practices and leading the way to a sustainable future by incorporating ESG values into their operations and policies. *Values at Work* offers the roadmap for investors who want to get to this future."

—Robert Jenkins, *Global Head of Research, Refinitiv/Lipper*

"More than a decade ago, Dan Esty led the way in teaching the capital markets how to drive corporate sustainability. His commercial sense and pragmatism helped us move from ideology to implementation. He and his team of authors in *Values at Work* are now bringing this same leadership to sustainable investing."

—Erika Karp, *Founder & CEO, Cornerstone Capital Group*

"Sustainable investing is not about financial products, it is about fulfilling the financial sector's fiduciary responsibility toward investors and asset owners. Professors Esty and Cort's book should get everyone speaking the common professional language that will enable the finance world to accelerate the transition toward a more sustainable global economic model."

—Patrick Odier, *Senior Managing Partner, Lombard Odier Group*

"Investors need information about what matters. If you are not asking for information about climate change, natural systems, how the workforce and community are treated, how power is used at the Board level, you are investing in a fantasy financial world or for another planet which we haven't got. *Values at Work* will get you on the right track."

—James Cameron, *Founder, Climate Change Capital*

"*Values at Work* is a masterful collection of great minds laying out a simultaneously practical and fascinating roadmap for the essential changes needed to improve sustainable investing and integrate sustainability into our capital markets."

—Mindy Lubber, *CEO and President, Ceres*

"Dan Esty and Todd Cort and their team of *Values at Work* authors have made a critical contribution to the market for sustainable investing. They underscore the urgent need for more accurate and quantitative ESG information to allow investors to make informed decisions on allocating capital."

—Matthew Arnold, *Global Head of ESG and Corporate Responsibility Engagement, JPMorgan Chase*

"*Values at Work* shows how financial regulations might be recast to ensure that capital flows to companies that are helping to address challenges such as climate change and sustainable development—and advancing transatlantic values more generally."

—Edouard-François de Lencquesaing, *President, European Institute of Financial Regulation*

Contents

Notes on Contributors

Christina Alfonso-Ercan advises top global fund managers and asset management firms across North America, Europe, and Latin America on sustainable investment strategies and ESG integration. A pioneer in using data-driven approaches to build high-performance ESG portfolios, she co-founded Madeira Global in 2012 to continue developing these insights. As Madeira's CEO, she is dedicated to elevating the mission of the financial markets to include impact-conscious practices. For this work, she was recognized as a "Game Changer" by the U.S. Green Chamber of Commerce, she has served on the advisory boards of PathNorth and the Milken Institute's Young Leaders Circle, and is a graduate of Fordham University and ESADE Business School.

Natalie Ambrosio Preudhomme is Director of Communications at Four Twenty Seven, where she manages publications, thought leadership, and engagements. Natalie's recent publications include, "Community Resilience and Adaptive Capacity: A Meaningful Investment Across Assets," published by the Federal Reserve Bank of San Francisco, and "Addressing Climate Risk in Financial Decision Making," in *Optimizing Community Infrastructure*. Previously, Natalie's work at the Notre Dame Global Adaptation Initiative helped to develop an assessment of U.S. cities' vulnerabilities to climate change and their readiness to adapt. Natalie holds a BS in Environmental Science and a certificate in Journalism, Ethics, and Democracy from the University of Notre Dame.

Jennifer Bender, Ph.D. is a Senior Managing Director at State Street Global Advisors (SSGA) and Director of Research for the Global Equity Beta Solutions team. Jenn joined SSGA in 2014 and since then has been responsible for leading SSGA's research across key areas of index investing, including core beta, smart beta, ESG, and thematic investing. Previously, Jenn was a Vice President in the Index and Analytics Research teams at MSCI and has held research roles at State Street Associates and Harvard University. Jenn also serves on the editorial boards for *The Journal of Portfolio Management* and *The Journal of Systematic Investing*. Jenn holds M.S. and Ph.D. degrees in Economics from Brandeis University.

Satyajit Bose is Professor of Practice at Columbia University, where he teaches sustainable investing, cost-benefit analysis, and mathematics. His research interests include the value of ESG information, carbon pricing, the link between portfolio investment and sustainable development in emerging markets, and the optimal use of environmental performance metrics for long horizon investment choices. He is co-author (with Dong Guo and Anne Simpson) of *The Financial Ecosystem: The Role of Finance in Achieving Sustainability* (Palgrave, 2019). He oversees a number of research projects at the Earth Institute's Research Program on Sustainability Policy and Management.

Todd Arthur Bridges is a Partner and Global Head of Sustainable Investing and ESG Research with Arabesque. His expertise lies at the intersection of sustainable finance, climate finance, market governance, and investment research. Todd has worked with global institutional investors across central banks, sovereign wealth funds, state pensions, corporate pensions, foundations, endowments, family offices, advisors, etc. Todd's education and training includes stints at Brown University, Harvard University, University of Chicago, and Cornell University in the United States as well as research fellowships at Oxford University and the Max Planck Society in Europe.

Todd Cort, Ph.D. works at the intersection of sustainability and investor value. For over 20 years, he has applied a scientific and economic lens to corporate social responsibility and environmental sustainability. Working with a diverse range of stakeholders, both through academia and consulting, Todd has sought to identify the areas that create the greatest value for investors, businesses, and society. Todd is a faculty member at the Yale School of Management with courtesy appointment at the Yale School of the Environment. He serves as Faculty Co-Director for the Yale Center for Business and the Environment (CBEY) and the Yale Initiative on Sustainable Finance (YISF). In addition, Todd serves as Treasurer on the board of Save the Sound, a regional environmental nonprofit serving communities around Long Island Sound.

Paul A. Davies is a Latham & Watkins partner in London who serves as the Local Department Chair of the Environment, Land & Resources Department and Co-Lead of Latham's Environment, Social & Governance Practice. He is recognized as a Thought Leader for Environment by *Who's Who Legal: Thought Leaders 2017–2019*, and was the only European lawyer named for the distinction from 2011 to 2017. He advises clients around the globe on ESG related matters. Paul holds an M.Phil. and LLB from the University of Wales, Cardiff.

Paul M. Dudek is a Latham & Watkins partner in Washington, D.C. who advises on all aspects of cross-border capital market transactions involving non-U.S. companies and sovereigns. He has particular insight on trends related to ESG reporting by public companies, as he previously served 26 years with the U.S. Securities and Exchange Commission's Division of Corporation Finance, most recently as Chief of the Office of International Corporate Finance. Paul holds a JD from New York University School of Law, cum laude, Order of the Coif, and a B.A. from Fordham University, Phi Beta Kappa, summa cum laude.

Michael Eckhart is Clinical Professor for Sustainable Finance at University of Maryland and Adjunct Professor at Columbia University. Previously, he was Global Head of Environmental Finance at Citigroup where he led the banking industry's development of the Green Bond Principles. Earlier, he was founding President of the American Council on Renewable Energy and President of the SolarBank Initiative in Europe, India, and South Africa, and had a 20-year career in power generation and renewable energy. He served in the U.S. Navy Submarine Service, and holds a BSEE degree from Purdue University and an M.B.A. from Harvard Business School.

Daniel C. Esty is the Hillhouse Professor at Yale University with primary appointments at the Yale School of the Environment and the Yale Law School and a secondary appointment at the Yale School of Management. He serves as Director of the Yale Center for Environmental Law and Policy and co-directs the Yale Initiative on Sustainable Finance. Dan is the author or editor of twelve books and dozens of articles on environmental and energy policy, corporate sustainability, and sustainability metrics. His recent (edited) volume, *A Better Planet: 40 Big Ideas for a Sustainable Future*, was named a top book of 2019 by *The Financial Times*. He served as head of the Connecticut Department of Energy and Environmental Protection from 2011 to 2014 and in several leadership roles at the U.S. Environmental Protection Agency from 1989 to 1993. He is a founder and principle of Constellation Research and Technology, a fintech firm focused on providing advanced sustainability data analytics to the investment marketplace.

Valerie S. Grant, CFA is a Senior Vice President and Senior Portfolio Manager at AllianceBernstein, where she manages the Responsible US Equities portfolio, co-chairs AB's Equity ESG Research Committee and serves on the Proxy and Governance Committee. Previously, she was a senior research analyst for several of AllianceBernstein's small-, mid- and large-cap equity portfolios. Valerie is a member of the Standards Advisory Group at the Sustainability Accounting Standards Board. She holds a BS in Economics from the Wharton School at the University of Pennsylvania and an MBA from Harvard Business School, where she was a George F. Baker Scholar.

Cary Krosinsky is an educator, author, and advisor with teaching that includes popular classes on sustainable finance at Brown and Yale. He is a leading author of several books on sustainable finance. His latest book focuses on modern China and calls for more cooperation to solve societal challenges. He is also Co-Founder of the Carbon Tracker Initiative and a member of the 2018 NY State Common Decarbonization Advisory Panel. He has written papers on a broad range of issues, including the Value of Everything, and in 2016, for the Principles for Responsible Investment (and their signatories) to aid the development of their asset owner climate change framework. His latest book is *Modern China: Financial Cooperation for Solving Sustainability Challenges* (Palgrave, 2020).

David A. Lubin is a serial entrepreneur and corporate sustainability thought leader. He has more than 30 years of business experience founding and managing business analytics companies including Spectrum Interactive, Palladium Group, and Renaissance Worldwide, a pioneer in enterprise performance management and tracking. David is a founder and managing director of Constellation Research and Technology, a sustainability data analytics firm. He previously served on the faculty at Tufts and Harvard University.

Emilie Mazzacurati is the founder and CEO of Four Twenty Seven (427mt. com), a leading provider of data on physical climate and environmental risks, and an affiliate of Moody's. Emilie has published extensively on climate risks disclosure, the impacts of climate change on equity and U.S. muni bond markets, and adaptation finance opportunities. Previously, Emilie was Head of Research at Thomson Reuters Point Carbon. She also served as an advisor to the Mayor of Paris on environmental policy. Emilie holds a Master of Political Science from the Institut d'Etudes Politiques de Paris and a Master of Public Policy from UC Berkeley.

Aisha I. Saad is a fellow at Harvard Law School's Program on Corporate Governance. She was previously a Research Scholar in Law and the Bartlett Research Fellow at Yale Law School. Aisha holds a J.D. from Yale Law School and a D.Phil. from Oxford University. Her research focuses on the public challenges of the modern corporation and legal doctrine relating to the corporate form. She has previously served as summer extern for Judge William Fletcher on the U.S. Court of Appeals for the Ninth Circuit and Judge Edward Chen on the U.S. District Court for the Northern District of California.

Kushal Shah is a Senior Associate at State Street Global Advisors (SSGA) in the Global Equity Beta Solutions Research team. He is responsible for research and development of equity strategies with a focus on incorporating ESG factors in the portfolio construction process. Prior to joining SSGA, Kushal worked at HDFC Bank with the FX Sales team. Kushal holds an M.S. in financial economics from Columbia Business School, an M.B.A. from the Indian Institute of Management, Bangalore, and a B.S. in Electrical & Electronics Engineering from the National Institute of Technology, Trichy.

Diane Strauss is the lead of the NGO Transport & Environment in France, where she seeks to accelerate the energy transition of the transportation industry through advocacy. She is the former Research Director of the Yale Initiative on Sustainable Finance (YISF) where she developed and conducted research projects. She has eight years of experience in sustainable finance and policy. Before joining YISF, Diane worked for the think tank 2DegreesInvesting Initiative, for the European policy office of the World Wildlife Fund (WWF) and for the French investment bank Natixis. She is the author of several research papers published with the UNEP-Inquiry, France Strategy, and Columbia University/Palgrave Macmillan editions, looking into the impact of public policies on the flow of capital for sustainable investments.

Hana V. Vizcarra is a staff attorney at Harvard Law's Environmental & Energy Law Program (EELP), where she leads EELP's portfolio on private sector approaches to climate and environmental issues. Her research includes analyzing legal issues in climate-related risk management and disclosure. Before joining EELP in 2018, Hana practiced environmental law for over seven years at two law firms. She represented clients in environmental litigation, counseled on compliance and regulation, and advised on environmental aspects of transactions. Before practicing, she worked in political research and communications for seven years. Hana received her JD from Georgetown Law and B.A. from Pomona College.

Ella Warshauer works at the intersection of business and environmental sustainability. She writes on sustainable finance and accelerating the transition to the low-carbon economy, and has researched and published papers on investing in green infrastructure, gender diversity, and the UN Sustainable Development Goals. Ella has experience in the financial services sector, having worked as an analyst at BlackRock, as well as in business operations, where she helped to scale a seed-stage tech startup. Ella earned her B.A. in Economics from Brown University.

Alison Weiner is an ESG Investment Strategist at State Street Global Advisors. Ali's career has focused on mobilizing the public, private, and nonprofit sectors to take collective action on social impact and responsibility goals. She previously worked as a Senior Strategist at Purpose, a strategy consultancy that builds social movements. There, she served businesses, foundations, and nonprofits, designing and implementing strategic advocacy campaigns on several health, human rights, and environmental issues. Ali holds an MBA from the Stanford Graduate School of Business, a Master of Public Administration from the Harvard Kennedy School, and BAs in History and Political Science from Yale College.

Kristina S. Wyatt serves as Latham & Watkins's Director of Sustainability in Washington, D.C., and has written extensively on ESG issues, particularly focusing on public company reporting, green finance, and board governance. She previously served as Legal Counsel to Commissioner Roel C. Campos of the U.S. Securities and Exchange Commission. Kristina holds an M.B.A. (Sustainability) from Yale University, a JD from the University of Colorado, and a B.A. from Duke University.

List of Figures

List of Tables

Part I

Introduction

1

Sustainable Investing at a Turning Point

Daniel C. Esty and Todd Cort

Abstract Sustainable investing has expanded from a niche interest to a mainstay of investment strategies around the world. With a growing number of investors focused not just on the financial promise of the companies in their portfolios but also the environmental, social, and governance performance of these enterprises, the demand for better ESG metrics and reporting has skyrocketed. This book explains the critical concepts, trends, risk frameworks, and investment tools that investors of all kinds—including those in stocks, bonds, private equity, infrastructure projects, and impact investing—need to know. With informative essays from a range of scholars, policy experts, and investment practitioners, it explores the state of play in sustainable finance with particular focus on the data, guidelines, legal standards, and principles required to make ESG reporting more trustworthy and thus sustainable investing more mainstream.

Keywords Sustainable investing · ESG reporting · Sustainability imperative · Financial regulation · Sustainable Development Goals · 2015 Paris Climate Change Agreement · Sustainable finance · Impact investing

D. C. Esty (✉) · T. Cort
Yale University, New Haven, CT, USA
e-mail: daniel.esty@yale.edu

T. Cort
e-mail: todd.cort@yale.edu

© The Author(s) 2020
D. C. Esty and T. Cort (eds.), *Values at Work*,
https://doi.org/10.1007/978-3-030-55613-6_1

Intro

Sustainable investing has become a booming investment category. According to the Morningstar 2020 Sustainable Landscape Report, there are now more than 300 sustainability-oriented mutual funds and Exchange Traded Funds (ETFs) in the United States with total assets as of 2020 just under $150 billion—up nearly 35% in the past three years.[1] While stock markets tumbled as a result of the COVID-19 pandemic, sustainability funds showed much greater resilience than other investment vehicles. And as markets recovered, the flow of capital into sustainable equity funds outpaced competing conventional equity funds.[2] As attention shifted from the public health crisis to economic recovery, a growing emphasis emerged in many circles on the need to "build back better"[3]—meaning more sustainably. In addition, the Black Lives Matter protests that erupted around the world in 2020 threw into high profile a series of social concerns including racial justice and economic inequality and raising questions about corporate cultures and workplace diversity. All of these recent events offer a signal of just how significant sustainable investing and finance have become—and explain why interest in *environmental, social, and governance* (ESG) metrics has taken off.

This interest spans the world. Indeed, the Global Sustainable Investment Alliance estimates that the assets under management (AUM) in funds that deploy some form of sustainability screening in their investment strategy now exceed $30 trillion.[4] While this figure may be an overestimate—because the depth of the sustainability focus considered by these funds varies widely—the scale of these numbers suggests that sustainable investing must be seen not as a fad but rather the new essential lens for investors in the twenty-first century.

Tracking Trends

The growing interest in sustainable investing reflects a number of trends that this book seeks to map and explain. Most notably, an ever-expanding number of mainstream investors have begun to insist that their portfolios better reflect their *values*—including their concerns across a spectrum of environmental, social, and governance (ESG) issues. Of course, investors range widely in their specific interest in ESG performance, with some focused narrowly on single issues such as climate change or diversity in the workplace. Others want to exclude companies that produce goods or services that they find objectionable, so some funds engage in *negative exclusions*—divesting from whole industry categories such as alcohol, tobacco, gambling, and firearms, or from poor performers on select issues such as greenhouse gas emissions or respect for human rights. And yet others simply want their investments tilted toward companies that are moving the world toward a sustainable future and to underweight those that are not helping to shift society onto a sustainable trajectory.

The rising interest in sustainable investing and finance, and the related focus on ESG reporting, has begun to reshape not just equity markets but the world of fixed income investments as well. So-called *green bonds*, for example, grew from $2.6 billion in 2012 to $257.7 billion in 2019.[5] As this volume demonstrates, a similar push toward sustainability can be found among private equity investors, hedge funds, and other specialized investment vehicles.

Investors vary not only in the sustainability issues they care about but also in how they deeply they want to lean into sustainability factors and their tolerance for risk, leading to a wide range of sustainable investing strategies. Some strategies seek to deliver outsized returns or *green alpha*. Others simply tilt toward sustainable companies using *smart beta* strategies. And some are aimed at investors who will accept sub-par returns because their focus is on *impact investing*, which means they prioritize societal benefits alongside their financial gains.

A second factor driving interest in the flow of capital toward support of a sustainable future can be traced to two landmark 2015 agreements: the UN Sustainable Development Goals (SDGs) and the Paris Climate Change Agreement. The 17 SDGs spell out a set of clear policy priorities for governments across the world, highlighting the need for improved results on a diverse set of critical challenges including hunger, poverty, clean water, economic development, human rights, and climate change.[6] Beneath the 17 topline goals, the UN effort specifies 169 quantitative targets to sharpen the focus on what needs to be done by not just governments, but also the business community and non-governmental organizations of all kinds. In this regard, the negotiations that led to both the 2015 Paris Agreement and the SDGs made it clear that success in achieving progress would require substantial flows of capital toward meeting the needs that had been defined.[7] Both agreements specifically indicate that traditional development assistance and other government funds will be insufficient to the task—and thus that private capital will be essential for expanded sustainability efforts in general and to the global response to climate change in particular.

Policymakers estimate that as much as $3 trillion per year in new investments will be needed to shift society toward a sustainable energy future and to meet the targets set by 2030 Sustainable Development Goals.[8] Delivering capital at this scale necessitates a vast increase in *sustainable finance*—with funds flowing toward sustainable projects, infrastructure, and companies. The UN Secretary General has outlined various barriers to the scale-up of funding for sustainability projects[9] with particular emphasis on "misaligned incentives and regulations, limited awareness, and difficulties in identifying, measuring, and reporting on sustainable investments."[10] Various chapters in *Values at Work* offer suggestions about how to overcome these obstacles.

Investment Logic

Both more data on corporate ESG performance and expanded incentives to channel capital toward companies and projects are needed to facilitate the transition to a sustainable future. But whether sustainable investing makes sense from a *financial* perspective continues to be debated. In the chapters that follow, we review the issues under contention. In particular, we explore the theoretical logic for investing in sustainability leaders, including the pathbreaking work of Harvard Business School professor Mike Porter, who hypothesizes that sustainability efforts can spur innovation.[11]

We also look at the empirical studies that purport to demonstrate a correlation between the integration of ESG indicators into portfolio choices and investment outperformance.[12] But while cutting-edge corporate environmental strategy and sustainability leadership leads to financial success in some instances,[13] not every claim of sustainability pays off. And the exact causal relationships and correlation circumstances remain unclear.[14] Moreover, some analysts theorize that companies focused on sustainability will underperform their peers, especially in markets that do not regulate effectively and permit pollution harms to go unaddressed (or as economists would say, for externalities to go *uninternalized*), thus damaging the competitiveness of companies that run out in front of the market in terms of their commitment to greenhouse gas emissions control or other sustainability actions.[15]

Some investors clearly anticipate a sweeping societal transformation toward a more sustainable future. They hope that a focus on ESG metrics will allow them to capture the upside of the economic shifts driven by this *sustainability imperative*.[16] They may seek to invest in renewable power companies or those selling energy efficiency goods and services that stand to benefit from the transition toward a clean energy future. Others will focus on divesting from companies that are heavily invested in fossil fuels that create *carbon exposure* and the risk of *stranded assets* such as oil and coal reserves that might never get to be exploited.

Yet other sustainability-minded investors see high scores on ESG metrics as a signal of resiliency and the capacity to outperform under difficult circumstances or in a down market. And indeed, the COVID-19 epidemic provides some evidence in support of this hypothesis, as sustainability funds weathered the market drop much better than their conventional counterparts.[17] In a world where resilient companies are more likely to deliver long-term growth, investors have shown increasing interest in ESG screening as an important signal of likely future marketplace success.

Ramping up Sustainable Investing

In this book, we review and explain the tools, standards, frameworks, and tangible efforts that will be needed to make ESG screening more effective and trustworthy—thus building the confidence of investors who want to bring a sustainability emphasis to their portfolios, which in turn will translate into scaled-up sustainable investing.[18] In Part II, we explore the current state of play in terms of measuring corporate ESG performance as well as the impact of sustainable investments. We highlight how ESG metrics will need to evolve and improve to allow greater comparability across companies—thereby permitting *real* sustainability leaders to be distinguished from those engaged in *greenwashing*—as well as to generate deeper insights for investors, providing ESG signals that are meaningful and likely to translate into long-term financial value.

In Part III, we look at the changing world of financial products in the sustainable investing marketplace. While sustainability-oriented fixed income products, such as green bonds, have scaled-up dramatically in recent years, bond funding will be insufficient to the scale of infrastructure that must be built. Thus, several chapters analyze how an ESG focus is being brought to other asset classes. In particular, we assess what will be required to expand equity investments, both public and private, to deliver the capital flows required for sustainable future growth.

In Part IV, we explore the complex and interdependent worlds of financial regulation, sustainability disclosure, and legal liability. ESG disclosures in the context of financial reporting represent one of the fastest changing elements of the sustainable investing realm, as investors demand increasingly detailed and decision-useful ESG performance data. Meanwhile, companies face difficulty in balancing potential legal challenges and business risks from disclosing too much ESG information with the legal liability that might emerge from failing to disclose *material* information on ESG-related risks. Likewise, regulators must balance the ESG needs and desires of investors against the possibility of being overly prescriptive or unduly burdensome in establishing corporate ESG disclosure rules that weigh companies down with heavy data production demands or create new information exposure risk.

We hope this book will provide the reader with a snapshot of the current state of practice in these areas critical to deployment of a new generation of methodologically rigorous and trusted ESG metrics positioned to undergird expanded investor confidence in sustainable investing. In addition, we hope that *Values at Work* will highlight how these practices, frameworks, regulations, and tools need to evolve to create a more mature sustainable finance arena. Given the momentum behind society's sustainability imperative, the ESG issues raised and shortcomings identified require urgent attention. By framing these needs, identifying the hurdles, and spelling out the array of parties that must be included in this process—such as governments, ESG data providers, financial managers, investors, auditors, corporate executives and boards, legal counsel, and NGO leaders—we hope that readers will see a role for themselves in delivering a more rigorous structure of ESG reporting and expanding the world of sustainable finance.

Notes

1. Morningstar. (2020). *Sustainable Funds U.S. Landscape Report.* https://www.morningstar.com/lp/sustainable-funds-landscape-report.
2. Shivaram, R. (2020, May 21). How Is ESG Performing In The Covid-19 Crisis? *Forbes.* Retrieved from https://www.forbes.com/sites/shivaramrajgopal/2020/05/21/how-is-esg-performing-in-the-covid-19-crisis/#6804908c4b62.
3. Furness, H. (2020, May 22). Prince Charles to Launch 'Great Reset' Project to Rebuild Planet in Wake of Coronavirus. *The Telegraph.* Retrieved from https://www.telegraph.co.uk/royal-family/2020/05/22/prince-charles-launch-great-reset-project-rebuild-planet-wake/; Mendiluce, M. (2020, April 3). COVID-19: How to Build Back Better with Climate Action. *World Economic Forum.* Retrieved from https://www.weforum.org/agenda/2020/04/how-to-build-back-better-after-covid-19/; World Resources Institute. (2020). Building Back Better After Coronavirus (COVID-19). Retrieved from https://www.wri.org/coronavirus-recovery.
4. Global Sustainable Investment Alliance. (2019). *Global Sustainable Investment Review.*
5. Climate Bonds Initiative. (2019) *Green Bond Market Summary.* https://www.climatebonds.net/resources/reports/2019-green-bond-market-summary.

6. United Nations. (2020). *Sustainable Development Goals.* https://sustainabledeve lopment.un.org/sdgs.

7. United Nations Conference on Trade and Development. (2019). *Financing a Global Green New Deal.* https://unctad.org/en/PublicationsLibrary/tdr 2019_en.pdf.

8. United Nations Conference on Trade and Development. (2019). *Financing a Global Green New Deal.* https://unctad.org/en/PublicationsLibrary/tdr 2019_en.pdf.

9. United Nations Secretary-General. (2019). *Roadmap for Financing the 2030 Agenda for Sustainable Development, 2019–2021.* https://www.un.org/sustai nabledevelopment/wp-content/uploads/2019/07/UN-SG-Roadmap-Financing- the-SDGs-July-2019.pdf.

10. United Nations Secretary-General. (2019). *Roadmap for Financing the 2030 Agenda for Sustainable Development, 2019–2021.*https://www.un.org/sustainab ledevelopment/wp-content/uploads/2019/07/EXEC.SUM_SG-Roadmap-Fin ancing-SDGs-July-2019.pdf.

11. Porter, M. & van der Linde, C. (1995, September–October). Green and Competitive: Ending the Stalemate. *Harvard Business Review.*

12. Friede, G., Busch, T., & Bassen, A. (2015). ESG and Financial Performance: Aggregated Evidence from More Than 2000 Empirical Studies. *Journal of Sustainable Finance & Investment, 5*(4), 210–233.

13. Esty, D., & Winston, A. (2009). Green to Gold: How Smart Companies Use Environmental Strategy to Innovate, Create Value, and Build Competitive Advantage.

14. Esty, D., & Cort, T. (2017). Corporate Sustainability Metrics: What Investors Need and Don't Get. *Journal of Environmental Investing, 54*(3), 140–154.

15. Esty, D., & Charnovitz, S. (2012, March). Green Rules to Drive Innovation. *Harvard Business Review;* Lyon, T. P., Delmas, M. A., Maxwell, J. W., Bansal, P., Chiroleu-Assouline, M., Crifo, P., … & Toffel, M. (2018). CSR Needs CPR: Corporate Sustainability and Politics. *California Management Review, 60*(4), 5–24; Vogel, D. J. (2005). Is There a Market for Virtue?: The Business Case for Corporate social Responsibility. *California Management Review, 47*(4), 19–45.

16. Lubin, D., & Esty, D. (2010, May). The Sustainability Imperative. *Harvard Business Review.*

17. Darbyshire, M. (2020, April 3). ESG Funds Continue to Outperform Wider Market. *Financial Times.* Retrieved from https://www.ft.com/content/ 46bb05a9-23b2-4958-888a-c3e614d75199.

18. For additional depth on the 'states of play' in ESG reporting, see Esty et al. (2020, September). *Toward Enhanced Sustainability Disclosure: Identi- fying Obstacles to Broader and more Actionable ESG Reporting.* Yale Initiative on Sustainable Finance White Paper.

Part II

Measuring Environmental and Social Impacts

2

Evolution of ESG Reporting Frameworks

Satyajit Bose

Abstract In response to increasing investor demand for non-financial information from companies, a number of sustainability accounting frameworks have evolved to improve standardized disclosure of environmental, social, and governance (ESG) information. These frameworks have created more consistent, readily available, and easily interpreted information for investors to assess the sustainability impact of capital allocation choices. The data that is easy to collect and disclose is, however, far less valuable than information that must be gleaned through complicated processes, extensive due diligence, collaborations with subject-matter experts, and serendipitous insights. ESG frameworks thus face a difficult trade-off between standardized information that is widely demanded and cheaply supplied versus nuanced and esoteric information required to form the basis of strategies capable of delivering market outperformance. Investors seeking *ESG-derived alpha* must thus look beyond these standardized data sources and frameworks for their deeper and more idiosyncratic analyses.

Keywords Global Reporting Initiative · Climate change exposure · Sustainable investing · UN Principles for Responsible Investment · ESG reporting · Carbon disclosure · Corporate sustainability strategy · ESG metrics · ESG-derived alpha · Impact investing · Impact reporting

S. Bose (✉)
Columbia University, New York City, NY, USA
e-mail: sgb2@columbia.edu

© The Author(s) 2020
D. C. Esty and T. Cort (eds.), *Values at Work*,
https://doi.org/10.1007/978-3-030-55613-6_2

Investors, along with a broad range of other stakeholders, increasingly demand disclosure of non-financial information beyond that which is currently available in financial statements. Many investors believe in the societal and private value of integrating environmental, social, and governance (ESG) considerations into financial decision-making as articulated by the UN Principles for Responsible Investment. Others harbor a narrower concern to generate financial outperformance through the pursuit of ESG *alpha*. In addition, modest pressure from some regulatory institutions to analyze the risks of climate change and extreme weather on corporate balance sheets has boosted investor interest in greater disclosure on the impact of global climate change trends on corporate assets and supply chains. There is thus considerable interest in revising accounting and disclosure frameworks to track measures of non-financial performance and incorporate analysis of climate change-related risks and opportunities.

To organize and render consistent the diversity of non-financial information potentially available, a number of sustainability accounting frameworks have evolved over the last quarter-century. Consistent, easily available, and easily interpreted ESG metrics are an essential requirement for any investor effort to measure the impact of capital allocation choices on the natural and social ecosystem. The work of sustainability accounting frameworks to render precision and inter-operability is vital to this task. Even the simplest efforts in this direction have facilitated flows of capital to low-carbon investments and sustainable development-linked funds. On the other hand, to the extent that asset managers integrate deeper sustainability-related knowledge into their pursuit of outperformance, information that is widely or easily known is often far less valuable than information that must be gleaned through complicated processes, extensive due diligence, collaborations with subject-matter experts, and serendipitous insights. It may be too much to expect that widely available and transparent frameworks provide the nuanced and esoteric information required to form the basis of persistent alpha isolation strategies. Investors are right to seek such information, but their search must necessarily lead them beyond the standardized data sources and frameworks.

Sustainability Reporting Frameworks: Logic of Standardization Versus Fragmentation

Methods of Organizing Information

Sustainability reporting frameworks provide a method of categorizing and regulating the semantics of non-financial information. The process of organization incorporates consensus-based typologies, definitions of concepts, controlled vocabularies, and methods of measurement. Frameworks are intended to advance precision, validity, consistency, and inter-operability. Most of the definitions and rules comprising sustainability accounting frameworks remain voluntary, lacking the force of government regulation. As such, sustainability frameworks have something in common with two common voluntary processes to regulate and mediate meaning: the promulgation of standards and the regulation of languages. Both of these activities mediate the varied interests of multiple stakeholders and aim to construct compromises and commonalities that will subsequently be upheld by most stakeholders.

The process of private, consensus-based standard-setting facilitates specialization, scale economies, and reduced transaction costs. In their history of standard-setting within the engineering profession, Yates and Murphy describe the timely and efficient process by which engineering associations and consensus-seeking committees of technical experts in Europe and North America were able to settle on common standards for such mundane but essential choices as screw thread characteristics and shipping container sizes.[1] Of course, the promulgation of standards does create a loss of diversity. For example, the domination of global standard-setting bodies by the United States after 1945 resulted in the elimination of the French and Austrian musical pitch of concert A at 435 hertz, due to the preference of American musicians for 440 hertz.[2] In any specific application, whether the value of scale economies exceed the value of the loss of diversity is an open question to be determined by the relative values of experimentation, resilience, and network externalities.

Natural languages present an example in which different approaches to standardization and inclusiveness exist in regulation. Natural languages tend to acquire new forms of expression and shed old ones in a decentralized manner. However, the official incorporation of words in dictionaries differs between languages, as exemplified by the much-discussed contrast between the English and French lexicographic methods.[3] The Oxford English Dictionary was designed to include everything: dialects, varieties, and the most obscure words from far-flung colonies.[4] By doing so, it provides a measure of legitimacy to even the least-used forms of expression. The *Dictionnaire de l'Académie Française,* on the other hand, is not meant to be encyclopedic, historical, or etymological. It is rather intended as a guide to modern usage, with new editions continually eliminating archaic words.[5] Both approaches validate the meanings of words, but the latter limits what is considered "good usage." Limiting the range of expression reduces the cost of communication across a network, but also limits what can be communicated. As we will discuss below, this trade-off also applies to the promulgation of reporting standards.

Is Framework Diversity a Weakness or a Strength?

In the context of sustainability accounting frameworks, a multiplicity of approaches to categorizing, defining, and expressing sustainability concepts have emerged. Some observers in the ESG investing community have voiced dissatisfaction at the presence of so many different and conflicting sustainability accounting frameworks. For example, Robert Eccles, a pioneering academic in the field of ESG and the first chairman of the Sustainability Accounting Standards Board (SASB), has stated that "With SASB, GRI [Global Reporting Initiative] and TCFD [Task Force on Climate-related Financial Disclosures], all offering different reporting standards, companies and investors have felt overwhelmed by the 'alphabet soup' of arbiters in the ESG industry."[6] In a similar vein, Gillian Tett opines that corporate pledges to address climate change "cannot be effective unless we put in place a commonly agreed system to track corporate exposures to climate risk—and right now this does not exist."[7] Framework diversity is generally more costly for the corporate issuers who must supply information than it is for the investors who consume such information. Corporate issuers refer to "reporting fatigue" resulting from the need to meet multiple demands for information.[8]

Conversely, there is an argument to be made for the benefits of diversity and the pitfalls of analytical monocultures in the evaluation of ESG performance.[9] In his bestseller, journalist James Surowiecki writes: "If one virtue of a decentralized economy is that it diffuses decision-making power (at least on a small scale) throughout the system, that virtue becomes meaningless if all the people with power are alike…or they become alike through imitation."[10] Many ESG ratings providers tout the range of underlying information sources as a strength of their ratings systems. For example, the rating provider CSRHub notes that it integrates information from 691 different sources in its ESG rating, including ESG analysts, government data, crowd-sourced information, and non-governmental organizations.[11]

Does the diversity of sustainability accounting frameworks pose an obstacle for investors that reduces the value of communication, or does it bolster the value of experimentation, resilience, and variety of analytical approaches? Is it better to have an "alphabet soup" of arbiters or a single dictatorial one? Clearly, the development of frameworks and standards is an iterative process that continually evolves. It remains perhaps too early to tell whether the frameworks for producing ESG disclosures will evolve into a single global standard with the precision and specificity of screw threads and container sizes or into the more decentralized and plastic guidelines represented by language usage, dialect, and idiom.

Frameworks for Non-Financial Reporting[12]

A broad range of frameworks comprise different typologies and categorizations of aspects of sustainability. A review of the major frameworks available to investors reveals that there is much collaboration among them, and very little duplication or contradiction. With few exceptions, they can be used in tandem. They all rely upon the Triple Bottom Line as a foundational conceptual framework for incorporating non-financial measures of performance into the evaluation of corporate activity.[13] John Elkington, the leading authority on corporate responsibility and sustainable development, introduced this concept, arguing that corporations should measure their net performance in the following three "bottom lines": the financial "profit and loss" account, the social "people" account, and the environmental "planet" account. The Triple Bottom Line represents perhaps the most widely accepted foundation for all the frameworks described below.

Global Reporting Initiative: Stakeholder Reporting

The most prevalent expression of a Triple Bottom Line framework for corporate reporting is the Global Reporting Initiative (GRI), which was founded in 1997 by the Coalition for Environmentally Responsible Economies, the UN Environment Program, and the Tellus Institute. In 2016, GRI launched its standards for sustainability reporting based on its 4th version of reporting guidelines launched in 2013. GRI standards are designed to guide the voluntary preparation of sustainability reports, which are generally published separately from regulatory filings. The GRI standards are routinely applied to specific disclosures and are not mutually exclusive to other frameworks listed herein. In principle, a firm can prepare a sustainability report according to other frameworks while still disclosing key performance indicators computed using GRI standards. GRI is by far the most widely adopted standard for sustainability reporting. Its database lists 7295 sustainability reports for 2017, of which 4202 (58%) were prepared according to GRI guidelines. The intended audience of the GRI consists of a broad range of stakeholders, including investors, consumers, employees, and civil society. GRI's mission is "to empower decisions that create social, environmental and economic benefits for everyone."[14] This focus on a range of stakeholders arises from GRI's origins within the United Nations dialogue around sustainable development, which does not privilege investors alone.

International Integrated Reporting Council: Integrated Reporting for Investors

The Integrated Reporting framework developed by the International Integrated Reporting Council (IIRC) aims to "improve the quality of information available to providers of financial capital to enable a more efficient and productive allocation of capital."[15] The IIRC explicitly targets providers of financial capital, while recognizing that there are multiple forms of capital. It aims to "enhance accountability and stewardship for the broad base of capitals (financial, manufactured, intellectual, human, social and relationship, and natural) and promote understanding of their interdependencies."[16] The IIRC was originally convened jointly in 2010 by the Accounting for Sustainability project of the Prince of Wales Charities and the GRI, drawing on the project's earlier work and on the King reports on corporate governance in South Africa.[17]

The IIRC framework proves far more difficult to apply than the GRI guidelines. The Integrated Reporting system is principle-based and requires a re-evaluation of the organization's business model, including how it creates value using the six types of capital outlined in the framework. This represents a radical re-thinking of the value creation narrative. The framework's articulation of the importance of six different types of capital is unique. Appropriately applied, the framework recognizes the importance of different stakeholders in the value creation process. However, the IIRC's 2013 standard has been criticized for its privileging of financial capital, its focus on providers of financial capital to listed companies, and its exclusion of context-based sustainability considerations.[18] Outside of South Africa, where the Johannesburg Stock Exchange has required Integrated Reporting for its listed companies since 2009, the adoption of the IIRC framework has lagged far behind that of GRI. The number of organizations using the IIRC framework for sustainability reporting amounts to approximately 15% of the number of organizations using the GRI framework.[19]

Sustainability Accounting Standards Board: Focus on Financial Materiality for Investors

In contrast to the Global Reporting Initiative (GRI) and in common with the International Integrated Reporting Council (IIRC), the Sustainability Accounting Standards Board (SASB) in the United States has adopted a focus on investors as the primary audience. The SASB Foundation was formed in 2011 by Jean Rogers, under the patronage of Michael Bloomberg, former Mayor of New York City and founder of the Bloomberg information service. The mission of the foundation is "to establish disclosure standards on sustainability matters that facilitate communication by companies to investors of decision-useful information."[20] In 2018, SASB issued 77 different standards covering the minimum sustainability reporting requirements for industries in 11 different sectors. As of 2020, 175 companies had prepared SASB-compliant sustainability reports.[21]

SASB emphasizes the notion of *financial materiality*, meaning that its standards focus on sustainability matters that are "reasonably likely to have a material impact on financial performance or condition."[22] SASB has chosen, through its multi-stakeholder process of standards creation, to identify the specific sustainability issues that are material to each of the 11 sectors for which it has issued standards. SASB has developed the "SASB Materiality Map" to codify its assessment of materiality by sector and issue.[23] The delineation of materiality by sector is based on a judgment about the relative importance to investors of business processes in specific sectors. For example, carbon emissions from fuel combustion is likely to be a more material issue for the transportation sector than for the financial sector. In this vein, according to the Map, business ethics is likely to be a material issue for more than 50% of industries in the financial and healthcare sectors, but less than 50% of industries in the extractives and mineral processing, infrastructure, resource transformation, services, and transportation sectors.[24] In determining which issues might be material to a given industry, SASB aims to ease the analytical burden for investors. The Materiality Map lightens the workload of an investment analyst, who can now point to an authority that relieves the duty to perform comprehensive due diligence on a full range of sustainability issues.

Impact Reporting Frameworks for Small and Medium Enterprises

The GRI, IIRC, and SASB frameworks are targeted toward large corporations, the majority of which are publicly traded. These reporting frameworks are most suited to large companies with access to an extensive reporting infrastructure, a range of specialized measurement professionals, and detailed academic research about the links between financial and sustainability performance. In the context of small and medium enterprises, the absence of extensive reporting resources has led to the development of somewhat more simplified reporting frameworks which nevertheless provide comparable measures of performance. Industry observers aiming to increase the level of standardization while reducing *reporting fatigue* are likely to find that these frameworks intended for smaller companies can provide useful examples of easy-to-compute metrics. The three frameworks utilized by a range of smaller companies and community organizations are the Impact Reporting and Investment Standards (IRIS), the B Impact Assessment, and the Future-Fit Assessment.

The Impact Reporting and Investment Standards are a catalog of performance metrics that many impact investors, primarily in private markets, use to measure social, environmental, and financial performance. The standards were designed for recipients of impact investing funds, which tend to be small companies, social enterprises, and community organizations. The standards were developed by the Global Impact Investing Network (GIIN) convened by the Rockefeller Foundation in 2009. In 2011, 29 impact investors signed a letter of support for this framework, committing to use it for the performance measurement of their own funds.[25] Since the impact investing community is significantly smaller than the global listed equity marketplace, an initiative supported by 29 organizations in this community carries significantly more weight than it would in the listed equity marketplace.

Two other frameworks, the B Impact Assessment and the Future-Fit Business Benchmark, are also designed for small and medium size enterprises. The B Impact Assessment is administered by the non-profit B Lab and is used to certify businesses into a network of "B Corporations."[26] Benefit corporations and businesses with stated social missions may use a B Lab certification to signal an objective and comprehensive rating of its impact on a range of stakeholders.[27] B Lab uses company-generated metrics as inputs in the computation of an impact assessment score that measures value creation for governance, workers, community, and environment. In order to be certified as a B Corp, a company must achieve an integrated score of 80 or above.[28] The

Future-Fit Benchmark is a publicly available guideline designed for firms to self-assess the fitness of their mission and operations for a systemically sustainable future.[29] The Benchmark is comprised of 23 social and environmental goals which the Future-Fit Foundation deems are minimum requirements for every business to ensure protection of people and planet.[30]

Climate Change-Related Frameworks

There are a number of frameworks for climate-related indicators, such as the Climate Disclosure Standards Board (CDSB), the Carbon Disclosure Protocol (CDP), and recommendations of the Task Force on Climate-related Financial Disclosures (TCFD). A key distinction should be drawn between efforts that aim to measure the environmental impact of corporate activity, such as the early versions of Carbon Disclosure Protocol, and efforts to measure the impact of changes in environment and climate on corporate balance sheets and financial performance, such as the Task Force on Climate-related Financial Disclosures. The former evaluates the impact of economic activity on ecosystems at large, while the latter evaluates the impact of changing ecosystems on the financial prospects of corporations.

Carbon Disclosure Protocol

The Carbon Disclosure Protocol (formerly known as the Carbon Disclosure Project) is a United Kingdom-based non-profit organization created in 2000 upon the initiative of a coalition of 35 institutional investors interested in using corporate carbon emissions data in their portfolio construction process. The Carbon Disclosure Protocol (CDP) sends questionnaires to the largest publicly traded companies regarding carbon emissions across their operations and supply chains and then compiles responses in a database that is available to the public and subscribers. This repository represents the longest-running time series of corporate climate change disclosures in existence. It has attempted to collect carbon emissions data on all Financial Times Global 500 firms since 2002, and all S&P 500 firms since 2006.[31] As of May 2020, the CDP's investor membership comprised 515 institutional investors with nearly $100 trillion in assets under management.[32] This level of investor representation makes it highly likely that companies receiving questionnaires will respond. Since the bulk of disclosures are responses to a common questionnaire, there is a high level of consistency across responses. As such, CDP data is widely used in academic studies of the relationship between environmental disclosure and financial performance. According to Matisoff et al., CDP reporting has contributed to significantly improved transparency for Scope 1 emissions, which stem from an organization's direct activities, and Scope 2 emissions, which represent indirect emissions from electricity use, since 2003.[33] However, the study found that transparency for Scope 3 emissions, which stem from supply chains, product lifecycles, and other indirect sources, remains lacking. There is also some evidence that corporate disclosures of carbon emissions in CDP surveys are more accurate and detailed than those in conventional corporate sustainability reports.[34] This finding demonstrates the value of CDP information relative to other disclosures.

CDP has partnered with the United Nations Global Compact (UNGC), World Resources Institute (WRI), and the World Wide Fund for Nature (WWF) to form the Science Based Targets initiative. The initiative provides a corporate emissions reduction target validation service.[35] Corporate targets are considered "science-based" if they are deemed consistent with the goals of the Paris Agreement to limit global warming to well-below 2 degrees Celsius above pre-industrial levels and pursue efforts to limit warming to 1.5 degrees Celsius.

Climate Disclosure Standards Board

The Climate Disclosure Standards Board (CDSB) is also a United Kingdom-based non-profit organization, founded at the World Economic Forum in 2007. The organization has published a framework and a set of principles that aim to help businesses report environmental and natural capital information with the same level of rigor that is customary for financial information and to encourage the reporting of consistent climate information that will help investors make decisions about strategy, investment performance, and future prospects.[36] The Standards Board has an explicit goal of constructing reporting principles based on other widely adopted standards and frameworks, such as Global Reporting Initiative and the CDP. CDSB has established seven principles for reporting environmental information, including a commitment to relevance, materiality, consistency, and comparability and an effort to be forward-looking. To this end, it is collaborating with Carbon Disclosure Project, Fujitsu, and the global standards consortium XBRL International to develop a climate change reporting taxonomy to facilitate consistency of reporting concepts across scales and geographies.[37] This initiative has the potential to eliminate some of the inconsistencies reported among corporate carbon emissions disclosures. In addition, CDSB has worked with the Task Force for Climate-related Financial Disclosures (TCFD) to launch a hub of information and tools related to climate change disclosure.

Task Force for Climate-Related Financial Disclosures

The Financial Stability Board (FSB) is a group of finance ministries and central banks from G20 countries established after the 2009 G20 summit in London. The Board is hosted and funded by the Bank of International Settlements in Switzerland. The FSB's Task Force for Climate-related Financial Disclosures (TCFD) is the most recent initiative in the context of corporate natural capital disclosure. The Task Force, formed in 2015, is comprised of 31 members selected by the FSB to broadly represent users and preparers of climate change disclosures. The TCFD was a response to a call by Mark Carney, then Chairman of the FSB:

> to develop climate-related disclosures that 'could promote more informed investment, credit, and insurance underwriting decisions' and, in turn, 'would enable stakeholders to understand better the concentrations of carbon-related assets in the financial sector and the financial system's exposures to climate-related risks.'[38]

As outlined in the quote, the TCFD embodies an effort to report on both impacts as well as dependencies on the environment. The TCFD aims to make climate change disclosures more actionable for investment banks, lenders, and insurance underwriters. In this sense, its target audience comprises an array of players who have intermediary roles in the investment ecosystem, rather than investors per se.

The TCFD is an initiative more directly linked to the financial sector than other previous endeavors. It is chaired by Michael Bloomberg, and its six-member secretariat consists of four Bloomberg employees. Almost all members of the task force are from for-profit reporting corporations, financial institutions, insurance companies, and key accounting or ratings providers in the financial ecosystem, with little representation from civil society. According to the TCFD website, as of 2020, more than 1027 organizations support the TCFD.[39] In a survey of 485 respondents from among these organizations, the TCFD found that 198 respondents were preparers of TCFD-recommended reports. Of these 198, 67% intend to implement the recommendations in the next 3 years.[40] Although the TCFD recommendations remain a voluntary framework with no jurisdictional authority, their association with the FSB gives them significant legitimacy within the financial sector because finance ministries and central banks can apply significant pressure and moral suasion to the financial ecosystem even in the absence of regulation. Nevertheless, the number of actual TCFD reports prepared by corporations will likely fall far short of the number of reports prepared in accordance with GRI standards.

The Sustainable Development Goals as a Sustainability Reporting Framework

In parallel with investor initiatives to broaden performance metrics of companies, the United Nations has pursued a process of expanding the quantitative measures of social and environmental performance, culminating in the development of the Sustainable Development Goals (SDGs) in 2015. The 2015 Goals encourage nations and private actors to conceptualize the triple bottom line of people, profit, and planet and strive for economic growth that balances social and economic development with environmental sustainability. The business sector was far more involved in the formulation of the SDGs than in the earlier Millennium Development Goals (MDGs), and many observers argue that the private sector can bring innovation, responsiveness, efficiency, and targeted skills to the achievement of the goals.[41] Investors directing corporate influence toward achieving SDG targets could have significant beneficial impact. In theory, the investor-driven capital allocation can go beyond creating jobs and fueling economic growth to take on more responsibility in promoting international sustainable development. To this end, the Global Reporting Initiative (GRI), the UN Global Compact, and the World Business Council for Sustainable Development have developed the SDG Compass, which provides a five-step approach to align business strategies with the SDGs: (1) understanding the SDGs; (2) defining priorities; (3) setting goals; (4) integrating; and (5) reporting and communicating.[42] Business toolkits, standards, and assessment frameworks from third-party organizations such as those described above can assist companies in achieving each of the 17 goals. The SDG Compass website lists 58 such tools, including the Aqueduct Water Risk Atlas, the Corporate Human Rights Benchmark, the Food Loss and Waste Protocol, the Global Protocol on Packaging Sustainability, the ISO 14000 family of standards on environmental responsibility, and the Bribe Payers Index.[43]

The Statistics Division of the Department of Economic and Social Affairs of the United Nations maintains a list of 231 official indicators that measure progress on the SDGs.[44] In addition, the Compass maintains an inventory of indicators produced by other organizations such as GRI or the

World Bank which align with specific SDGs. Although investors are not the primary intended audience for SDG indicator information, the concept of the SDGs has proved popular as a feature of investment products. A number of SDG-linked bonds and SDG-aligned investment funds have been launched.

Some Differentiating Attributes of Frameworks

Although there is much in common between the different ingredients of the "alphabet soup," the frameworks discussed above do differ in emphases. We outline below some key points of divergence.

Stakeholders vs. Investors

A key differentiator between the Global Reporting Initiative (GRI) and Sustainable Accounting Standards Board (SASB) is the former's focus on a broad range of stakeholders vs. the latter's target audience of investors. In a joint article in 2017, the CEOs of GRI and SASB write: "Rather than being in competition, GRI and SASB are designed to fulfill different purposes for different audiences. For companies, it's about choosing the right tool for the job."[45]

GRI's recent Standard 206 on disclosure on Tax and Payments to Governments illustrates its applicability for a broad range of stakeholders. In a detailed account of the available methods of valuing intellectual capital and allocating it to tax-advantaged jurisdictions, Wiederhold demonstrates how multinational companies can drastically and legally reduce their tax burden.[46] GRI's standard requires public country-by-country reporting of taxes paid by a multinational corporation. Such disclosure would sharply increase transparency for taxing jurisdictions and has the potential to discourage aggressive tax avoidance. In theory, tax avoidance benefits shareholders and certain accounting and tax advisory professionals, but adversely affects the funds available for public infrastructure and social welfare, hurting almost all other stakeholders. Non-shareholder stakeholders have a collective incentive to minimize the adverse effects of aggressive tax avoidance, since they bear a significant share of the burden of externalities that cannot be ameliorated due to precarious government financing. It is perhaps not a coincidence that the shareholder-focused accounting frameworks (IIRC and SASB) have so far failed to propose any standard on tax transparency.

While it is true that an investor may not care in the short term if a company aggressively avoids taxes, in the long run, it is not in the interest of that investor to ignore signals of future pressure against tax-dodging practices. Investors like to be informed. Information intended for other stakeholders is important for investors to know. Hence, GRI standard disclosures are perhaps just as relevant as more targeted SASB disclosures for the investor who is interested in the long-term sustainability of return.

The Meaning of Materiality

SASB differentiates itself from other frameworks through its avowed attempt to codify materiality. Research suggests that corporate managers' focus on material issues by sector increases the value-relevance of their sustainability investments.[47] However, in a world of disruptive change, the classification of certain issues as negligible and others as critical through a cumbersome and occasional process of standard-setting will produce an inaccurate assessment of materiality. For example, in its Materiality Map, SASB determined that business ethics issues are not likely to be material for the technology and communications sector, despite much evidence to the contrary.[48] Similarly, an asset manager might be surprised to learn that SASB does not consider data security or customer privacy to be material sustainability issues for the asset management industry. SASB considers data security to be a material issue for commercial banks, but not customer privacy. In 2019, Capital One, a commercial bank and credit card company, announced that its data concerning 100 million U.S. citizens and 6 million Canadian residents had been stolen by a hacker.[49] While this was clearly a data security issue, it also had adverse impacts on customer privacy. An investor comparing portfolio holdings in Capital One with its competitors would likely care to perform due diligence on Capital One's processes to protect customer privacy. These examples demonstrate that it is a fool's errand to predetermine the types of issues which will be material to an investor in the way SASB purports to. Investors cannot allow themselves the luxury of outsourcing the definition of materiality to a static process administered by a standards-setting body. For example, Carol Adams of Durham University Business School points out:

> An investor would be wise not to ignore a narrative disclosure that farmers in a drought-stricken area are complaining to politicians that they need the water used by a bottling plant more. That sort of information might only be gleaned through a company's process of engaging with stakeholders—a process

that the Global Reporting Initiative Standards are unique in requiring them to disclose.[50]

Incommensurable Climate Scenario Analysis

Both GRI and SASB aim to make disclosed key performance indicators comparable across companies. In the case of TCFD, there is little focus on cross-issuer comparability. The TCFD reports that users of climate-related financial disclosures require companies to provide more clarity on the potential financial impact of climate-related issues on their business prospects.[51] TCFD recommends scenario analysis by report preparers, but currently there is very little common guidance on the parameters and assumptions underlying such scenarios. Although the Intergovernmental Panel on Climate Change (IPCC) and the International Energy Agency (IEA) have developed policy-relevant scenario descriptions, there are no "standard scenarios" that incorporate climate change impacts at local scale, climate-related drivers of business performance, and parameters of climate change uncertainty related to business planning assumptions. This makes it impossible to compare TCFD-recommended disclosures across companies, severely limiting the utility of such disclosures for investors. The TCFD cannot be considered a standard in the sense of GRI or SASB. It remains all too easy for companies reporting within this framework to consider idiosyncratic climate risks that are not fully described to investors and then conclude that their business models are resilient to such risks. For example, out of four illustrative company reports highlighted in TCFD's latest status report, all four declare that their strategies are "resilient" or "robust" to climate risks.[52]

Why Do Investors Need Information and Can Frameworks Provide It?

One reasonable interpretation of sustainability for the investor is the identification and protection of the sources of repeatable flows of benefits. For the long-term, especially universal investor, a holistic understanding of the drivers of the sustainability of cash flows implies integrating reporting on the levels of natural and human capital that facilitate long-term returns. There is a commonality between the calls of civil society for sustainable development, such as the balancing of economic growth and environmental and social objectives, and the interest of the universal investor in avoiding companies that generate short-term returns from the unsustainable liquidation of natural and human capital. Furthermore, the widespread recognition of the advent of the citizen investor and dispersed ownership suggests that viewing investors as primarily owners of and lenders to companies is a woefully incomplete description. Today, the investor constituency now intersects far more broadly than in previous centuries with the consumer, employee, and community constituencies. Therefore, a neat separation of the interests of investors and the broader cross-section of stakeholders in the focus of frameworks for sustainability measurement is neither productive nor possible.

At the beginning of this chapter, we outlined two reasons investors may care about ESG information: (1) to assess the impact of corporate activity on environmental and social systems and (2) to identify sources of ESG "alpha." In service of the first reason, consistent, easily available, and easily interpreted metrics are essential. For this purpose, there is no doubt that standardization and centralized definitions are critical. The establishment of a consistent method of determining Scope 1 and 2 emissions is a prerequisite for determining whether corporate efforts at emission reduction are effective.[53] Even relatively simple pieces of information can be incredibly helpful for responsible investor efforts.

Information is also the primary raw material in the process of generating value through a process of security selection, or pursuing ESG "alpha." In this activity, information that is widely or easily known is far less valuable than information that must be gleaned through complicated processes and extensive due diligence combined with serendipitous insights. To the extent that information is widely available and similarly interpreted, it is likely to be incorporated into asset prices and therefore unlikely to lead to outperformance. An implication of most versions of the efficient market hypothesis is that such information is not worth its costly acquisition. Obscure, hard-to-interpret information, on the other hand, may be valuable for security selection. Widely available and transparent frameworks cannot provide the esoteric information needed to form the basis of alpha isolation strategies. If they could, the low barriers to entry would quickly dissipate any outperformance, obviating the original impetus to gather such arcane information. Investors do seek such information, but their search must necessarily lead them beyond the standardized data sources. The search for arcane but decision-relevant information is exemplified by the alliances forged between active investment managers, Alliance Bernstein and Wellington Management, and climate science institutions, Lamont Doherty Earth Observatory and Woods Hole Research Center, respectively. These alliances have the potential to radically upgrade capital allocation towards eliminating climate risk and funding solutions.[54]

The financial ecosystem needs both types of information: (a) clear and consistent, and (b) inchoate and arcane. Sustainability reporting frameworks can help with the former. They are likely to have little to contribute to the latter.

What Is Next in the Evolution of Frameworks?

While indicators of sustainability at the issuer level have the potential to improve capital allocation choices for investors, they also provide important metrics of success and intermediate progress for a broader range of stakeholders focused on universal sustainable development. Quantitative metrics of sustainability can transform a vast amount of information about our complex environment into concise, policy-applicable, and decision-relevant information. We have described a number of structured frameworks that limit and categorize the universe of metrics. There is some likelihood of coalescence between frameworks in the near future, although the investor-focused and stakeholder-focused approaches do not share a common vision.

It remains unlikely that a single global standard such as that for screw threads or container sizes will ever dominate the provision of ESG information. It is clear today that the Global Reporting Initiative (GRI) is a far more widely adopted framework than the Sustainability Accounting Standards Board (SASB). SASB's focus on the investor will undoubtedly limit the possibility of garnering the universal legitimacy that a stakeholder-focused initiative, such as GRI, can aspire to. Furthermore, if the geopolitical tensions between the United States and its allies vs. China and Russia observed in the wake of the coronavirus pandemic persist, then the U.S.-centric nature of SASB's approach will diminish its global appeal. Nevertheless, as we have noted earlier, all the frameworks described bring their own foci to bear on the problem of integrating a broad range of ESG metrics into the capital allocation process.

Investors can view the multiplicity of frameworks and metrics as an obstacle or an opportunity. While some bemoan the lack of standardization in frameworks and the consequent leeway in the measurement of sustainability performance, the diversity in investor goals and the shroud of incomplete information underline the value of diverse approaches and experimentation to capturing value through security selection. Adaptability and evolution of frameworks and composite indexes of sustainability is essential because the collective understanding of sustainability remains in flux. Investor motivations are sufficiently diverse that there remain important roles both for simplified, consistent frameworks and also for in-depth, arcane, and hard-to-interpret information.

Notes

1. Yates, J., & Murphy, C. N. (2019). *Engineering Rules: Global Standard Setting Since 1880*. Baltimore, MD: Johns Hopkins University Press.
2. Immerwahr, D. (2019). *How to Hide an Empire: A History of the Greater United States* (1st ed.). New York: Farrar, Straus & Giroux.
3. Estival, D., & Pennycook, A. (2011). L'Academie Francaise and Anglophone Language Ideologies. *Language Policy, 10*(4), 325–341. Retrieved from https://link.springer.com/article/10.1007/s10993-011-9215-6.
4. Winchester, S. (2003). *The Meaning of Everything: The Story of the Oxford English Dictionary*. Oxford, UK: Oxford University Press.
5. Estival, D., & Pennycook, A. (2011). L'Academie Francaise and Anglophone Language Ideologies. *Language Policy, 10*(4), 325–341.
6. Temple-West, P. (2019, October 6). Companies Struggle to Digest 'Alphabet Soup' of ESG Arbiters. *Financial Times*. Retrieved from https://www.ft.com/content/b9bdd50c-f669-3f9c-a5f4-c2cf531a35b5.
7. Tett, G. (2020, January 16). The Alphabet Soup of Green Standards Needs a New Recipe. *Financial Times*. Retrieved from https://www.ft.com/content/b3fadc18-3851-11ea-a6d3-9a26f8c3cba4.
8. Pavoni, S. (2020, May 11). Proliferation of Demands Risks 'Sustainability Reporting Fatigue'. *Financial Times*. Retrieved from https://www.ft.com/content/9692adda-5d73-11ea-ac5e-df00963c20e6.
9. See pages 114–120 in Bose, S., Guo, D., & Simpson, A. (2019). *The Financial Ecosystem: The Role of Finance in Achieving Sustainability*. London, UK: Palgrave Macmillan.
10. Surowiecki, J. (2004). *The Wisdom of Crowds: Why the Many Are Smarter Than the Few and How Collective Wisdom Shapes Business, Economies, Societies, and Nations*. New York: Doubleday.
11. CSRHub. (2020). *CSRHub: Sustainability Management Tools*. Retrieved from https://www.csrhub.com/.
12. This section draws heavily upon and updates material presented in pp. 94–104 of Bose, S., Guo, D., & Simpson, A. (2019). *The Financial Ecosystem: The Role of Finance in Achieving Sustainability*. London, UK: Palgrave Macmillan.
13. Elkington, J. (1998). Partnerships from Cannibals with Forks: The Triple Bottom Line of 21st-Century Business. *Environmental Quality Management, 8*(1), 37–51.
14. Global Reporting Initiative. (2019). *GRI*. Retrieved from https://www.globalreporting.org/Pages/default.aspx.
15. International Integrated Reporting Council. (2013). *The International <IR> Framework*. Retrieved from http://integratedreporting.org/wp-content/uploads/2015/03/13-12-08-THE-INTERNATIONAL-IR-FRAMEWORK-2-1.pdf.
16. International Integrated Reporting Council. (2013). *The International <IR> Framework*.

17. The Prince's Accounting for Sustainability Project & Global Reporting Initiative. (2010). *Press Release: Formation of the International Integrated Reporting Committee.* Retrieved from https://integratedreporting.org/wp-con tent/uploads/2011/03/Press-Release1.pdf.
18. McElroy, M. W. (2017). With the Changing of the Guard at the IIRC, a Challenge to Richard Howitt. *Sustainable Brands.* Retrieved from https://sustai nablebrands.com/read/new-metrics/with-the-changing-of-the-guard-at-the-iirc-a-challenge-to-richard-howitt.
19. Bose, S., Guo, D., & Simpson, A. (2019). *The Financial Ecosystem: The Role of Finance in Achieving Sustainability* (pp. 96). London, UK: Palgrave Macmillan.
20. Sustainability Accounting Standards Board. (2018). Mission. *SASB.* Retrieved from https://www.sasb.org/governance/.
21. Sustainability Accounting Standards Board. (2020). Companies Reporting with SASB Standards. *SASB.* Retrieved from https://www.sasb.org/company-use/sasb-reporters/.
22. Sustainability Accounting Standards Board. (2018). Mission. *SASB.*
23. Sustainability Accounting Standards Board. (2018). SASB Materiality Map. *SASB.* Retrieved from https://materiality.sasb.org/.
24. Sustainability Accounting Standards Board. (2018). SASB Materiality Map. *SASB.*
25. Global Impact Investor Network. (2011). *Twenty-nine Impact Investors Sign Letter of Support for IRIS Initiative.* Retrieved from https://thegiin.org/assets/binary-data/MEDIA/pdf/000/000/19-1.pdf.
26. B Lab. (2020). About B Corps. *Certified B Corporation.* Retrieved from https://bcorporation.net/about-b-corps.
27. Cho, M. (2017). Benefit Corporations in the United States and Community Interest Companies in the United Kingdom: Does Social Enterprise Actually Work? *Northwestern Journal of International Law & Business, 37*(1), 149–172.
28. Nigri, G., & Del Baldo, M. (2018). Sustainability Reporting and Performance Measurement Systems: How Do Small- and Medium-Sized Benefit Corporations Manage Integration? *Sustainability, 10*(12).
29. Future-Fit Foundation. (2019). *Future-Fit Business.* Retrieved from https://futurefitbusiness.org/.
30. Kendall, G., & Rich, M. (2019). The Future-Fit Business Benchmark: Flourishing Business in a Truly Sustainable Future. In J. Walker, A. Pekmezovic, & G. Walker (Eds.), *Sustainable Development Goals: Harnessing Business to Achieve the SDGs Through Finance, Technology, and Law Reform* (pp. 235–252). Hoboken, NJ: Wiley.
31. Kim, J. (2018). *A Multi-Step Model of Boundary Spanning and Absorptive Capacity: The Differential Impact of Board and Top Management Team Experience on the Development of Sustainability-Related Capabilities* (Doctoral dissertation, Arizona State University). ProQuest Information & Learning.
32. CDP. (2020). What We Do. *CDP.* Retrieved from https://www.cdp.net/en/info/about-us/what-we-do.

33. Matisoff, D. C., Noonan, D. S., & O'Brien, J. J. (2013). Convergence in Environmental Reporting: Assessing the Carbon Disclosure Project. *Business Strategy & the Environment, 22*, 285–305. https://doi.org/10.1002/bse.1741.

34. Depoers, F., Jeanjean, T., & Jérôme, T. (2016). Voluntary Disclosure of Greenhouse Gas Emissions: Contrasting the Carbon Disclosure Project and Corporate Reports. *Journal of Business Ethics, 134*(3), 445–461. Retrieved from https://www.jstor.org/stable/24703782?seq=1#metadata_info_tab_contents.

35. Science Based Targets. (2020). About the Science Based Targets Initiative. *Science Based Targets.* Retrieved from https://sciencebasedtargets.org/about-the-science-based-targets-initiative/.

36. Climate Disclosure Standards Board. (2020). Framework for Environmental and Climate Change Information. *Climate Disclosure Standards Board.* Retrieved from https://www.cdsb.net/what-we-do/reporting-frameworks/enviro nmental-information-natural-capital.

37. Climate Disclosure Standards Board. (2020). XBRL Project Governance. *Climate Disclosure Standards Board.* Retrieved from https://www.cdsb.net/pri orities/xbrl/xbrl-project-governance.

38. Task Force on Climate-Related Financial Disclosures. (2017). *Final Report: Recommendations of the Task Force on Climate-Related Financial Disclosures.* Retrieved from https://www.fsb-tcfd.org/wp-content/uploads/2017/06/FINAL-TCFD-Report-062817.pdf.

39. Task Force on Climate-Related Financial Disclosures. (2020). *Task Force on Climate-Related Financial Disclosures.* Retrieved from https://www.fsb-tcfd.org/.

40. Task Force on Climate-Related Financial Disclosures. (2019). *2019 Status Report.* Retrieved from https://www.fsb-tcfd.org/wp-content/uploads/2019/06/2019-TCFD-Status-Report-FINAL-053119.pdf.

41. Scheyvens, R., Banks, G., & Hughes, E. (2016). The Private Sector and the SDGs: The Need to Move Beyond 'Business as Usual'. *Sustainable Development, 24*, 371–382. Retrieved from https://doi.org/10.1002/sd.1623.

42. GRI, UN Global Compact, & WBCSD. (2015). *SDG Compass: The Guide for Business Action on the SDGs.* Retrieved from https://sdgcompass.org/wp-con tent/uploads/2015/12/019104_SDG_Compass_Guide_2015.pdf.

43. GRI, UN Global Compact, & WBCSD. (2015). Inventory of Business Tools. *SDG Compass.* Retrieved from https://sdgcompass.org/business-tools/.

44. United Nations. (2020). SDG Indicators. *Sustainable Development Goals.* Retrieved from https://unstats.un.org/sdgs/indicators/indicators-list/.

45. Mohin, T., & Rogers, J. (2017). How to Approach Corporate Sustainability Reporting in 2017. *GreenBiz.* Retrieved from https://www.greenbiz.com/art icle/how-approach-corporate-sustainability-reporting-2017.

46. Wiederhold, G. (2014). *Valuing Intellectual Capital: Multinationals and Taxhavens.* New York: Springer Science.

47. Khan, M., Serafeim, G., & Yoon, A. (2016). Corporate Sustainability: First Evidence on Materiality. *Accounting Review, 91*(6), 1697–1724. Retrieved from https://doi.org/10.2308/accr-51383

48. Bose, S., Guo, D., & Simpson, A. (2019). *The Financial Ecosystem: The Role of Finance in Achieving Sustainability* (pp. 98). London, UK: Palgrave Macmillan.

49. McLean, R. (2019, July 30). A Hacker Gained Access to 100 Million Capital One Credit Card Applications and Accounts. *CNN Business.* Retrieved from https://www.cnn.com/2019/07/29/business/capital-one-data-breach/index.html.

50. Adams, C. A. (2020, January 23). Investors Are Asking the Wrong Questions on Sustainability. *Financial Times.* Retrieved from https://www.ft.com/content/7f9fd437-30fd-4f0a-9aa1-faa61dcbb399.

51. Task Force on Climate-Related Financial Disclosures. (2019). *2019 Status Report.*

52. Task Force on Climate-Related Financial Disclosures. (2019). *2019 Status Report.*

53. Temple-West, P. (2019, October 6). Companies Struggle to Digest 'Alphabet Soup' of ESG Arbiters. *Financial Times*; Tett, G. (2020, January 16). The Alphabet Soup of Green Standards Needs a New Recipe. *Financial Times.*

54. AllianceBernstein. (2019). *Collaboration with Columbia on Climate Change and Investing.* Retrieved from http://www.alliancebernstein.com/abcom/email/retail/2019/documents/columbia-climate-collaboration.pdf; McGlinchey, D. (2018). *Wellington Management and Woods Hole Research Center Announce Strategic Climate Science Initiative.* Woods Hole Research Center. Retrieved from https://whrc.org/wellington-management-and-woods-hole-research-center-announce-initiative/.

3

ESG Risk Depends on Management Control Quality

Todd Cort

Abstract Interest from investors in environmental, social, and governance (ESG) data has grown in part because of a string of corporate mis-steps associated with social and environmental impacts that have resulted in substantial losses for investors. These mis-steps clearly demonstrate that our current methods to measure ESG performance fail to capture the nature of many risks faced by companies. Nor do they gauge corporate resilience and the capacity of companies to mitigate the financial impacts of ESG-related risks. This chapter argues that investors need better corporate governance metrics that capture these sorts of environmental and social management challenges and signal how these issues will be internalized within the finances of a company. It further proposes a set of more granular metrics that look specifically at the ability of a company to mitigate environmental and social risk through more robust governance structures and management controls. Understanding whether governance systems are in place and positioned to identify and mitigate environmental and social risks (G^{ES}) will be important for investors seeking to avoid the next corporate mis-step and investing for long-term financial value.

Keywords ESG performance · ESG data · Risk management · Financial performance · Climate change · Sustainable investing · ESG metrics · Materiality · Good governance · ESG-related risks

T. Cort (✉)
Yale University, New Haven, CT, USA
e-mail: todd.cort@yale.edu

© The Author(s) 2020
D. C. Esty and T. Cort (eds.), *Values at Work*,
https://doi.org/10.1007/978-3-030-55613-6_3

Investor focus on corporate responsibility has grown substantially over the last decade, with the past few years seeing a particular spike in interest in sustainability.[1] Thus far, much of the focus has been on environmental, social, and governance (ESG) performance data. However, the structures within companies to manage risk and govern strategy clearly play an important role as to whether this ESG performance will result in financial impact or gain. Therefore, it is important to move from a framework emphasizing ESG, to a more nuanced evaluation of *governance, as it is used to manage environmental and social impact* (G^{ES}). In this chapter, three examples are analyzed where a lack of clear governance structures to manage environmental and social risks created massive financial consequences.

The pace of growth and the breadth of interest reflect a growing body of research suggesting positive correlations between corporate sustainability and financial performance.[2] However, the wide variety of factors in ESG means it is currently unclear which factors (or combination of factors) underlie the improved financial performance. As a result, the predominant conclusion from investors is that long-term financial performance can likely be enhanced by integrating ESG performance metrics into investment decision-making, but ensuring that a wide variety of ESG metrics are included in the assessment in order to "hedge their bets." In response, there has been an increasingly wide range of ESG performance data disclosed by companies in both public and private equity as well as fixed income instruments.

A close look at the empirical studies suggests a number of factors that may drive the observed correlations. Management practices, including the presence of policies, controls, and ESG-aligned business strategies, lead the list of potential drivers. Research suggests that *High Sustainability* companies—companies with policies and management systems for environmental responsibility—outperform their peers in both stock market and accounting results.[3] Similar positive correlations have been found between financial performance and management practices on specific material ESG issues within sectors.[4] Reports of companies with specific ESG management practices performing better financially have likewise emerged from the practitioner literature.[5]

Although the correlation between long-term financial performance and stronger ESG performance data appears to be solidifying, the cause remains unclear. For example, we do not yet know whether better ESG performance drives financial performance or that financial performance and ESG performance are both a function of better managed companies. Critics argue that the dominant determinant of financial performance is management quality and practices—and that all other factors pale in comparison—concluding

that ESG performance data and financial performance have no causal relationship and are both determined independently by good or bad management
practices.[6]

The Failure of ESG Performance Data

Much of investors' ESG data collection over the last four years has been
targeted at identifying risks. Risk identification generates information that
is useful for both the reporting company and investors. However, the actual
information that companies are encouraged or required to report on ESG
issues tends to be wide-ranging and not standardized, which makes them
less useful. Through wide stakeholder involvement, including with investors,
organizations—such as the Global Reporting Initiative (GRI), Sustainability
Accounting Standards Board (SASB), International Integrated Reporting
Council (IR), and Climate Disclosure Standards Board (CDSB)—have developed guidance for reporting ESG information through mainstream reporting
channels alongside financial information. Despite these efforts, research
reveals significant challenges in achieving integration of decision-useful ESG
information into existing mainstream financial reports.[7] In other words, we
are still having trouble measuring and reporting information that properly
reflects financial risks and opportunities resulting from environmental or
social factors.

Two recent efforts to create more decision-useful information for investors
bear special mention. First, the Task Force on Climate-related Financial
Disclosure (TCFD), spearheaded by Mark Carney of the Bank of England,
asks companies to assess their exposure to the various risks posed by climate
change. This might include risks from new regulation on carbon emissions
(mitigation risks), risks from severe weather, drought, or other business
disruptions (adaptation risks), or even risks from loss of market access as
energy, infrastructure, and other economic segments change (transition risks).
The TCFD asks companies to assess these risks under one or more future
scenarios so that investors can take a view on the likelihood of the risk and
the risk-mitigation efforts of the company. To date, disclosures by companies referencing the TCFD Guidelines have been inconsistent, especially with
regard to the financial implications of climate change risk and the business
strategies undertaken to address these risks.[8]

The second significant effort centers on the alignment of financial risk
with ESG materiality assessments. The Committee of Sponsoring Organizations of the Treadway Commission (COSO)[9] drafted an *Internal Control*

Framework demonstrating a pathway by which financial auditing could be applied to environmental and social performance data to improve the reliability (i.e., decision-usefulness) of information.[10] In addition, the World Business Council for Sustainable Development (WBCSD), in partnership with Committee of Sponsoring Organizations of the Treadway Commission and others, released guidance on the application of enterprise risk management (ERM) processes to the assessment and compilation of ESG risks.[11] While the sustainable investing world is not yet at the point of full agreement on these methodological approaches, the two documents lay out a set of tools that companies can use to better identify ESG data used for disclosure or assessment of risks—and thus represent a step forward for those seeking consistency in ESG data. In short, we are getting closer to having data that is both comparable and accurate to support better decision-making.

Materiality is an important concept to identify environmental or social risks that may be increasing as a result of global mega-trends or changes in stakeholder perceptions. Both the Task Force on Climate-related Financial Disclosure and Committee of Sponsoring Organizations of the Treadway Commission efforts are predicated on answering the question: "what environmental and social issues are *material* to performance?" The principle of *materiality*, as defined by the Global Reporting Initiative, posits that individual companies and their stakeholders are best positioned to determine which issues (and associated data) are most important. The Sustainability Accounting Standards Board takes a similar view, but narrows the scope of stakeholders down to investors and therefore asks "what environmental and social aspects are *financially material*?"[12] As such, much of the efforts by companies and investors to date have been to understand the potentially material environmental and social aspects faced by companies and to measure performance of the company against these aspects. For example, climate change is a material environmental risk for all sectors, and therefore, we have seen a surge in the amount of data on greenhouse gas emissions.

However, despite our increasingly accurate understanding of the material issues based on today's ESG data, decision-useful information and associated metrics to predict corporate risks related to ESG remain elusive. A few examples illustrate the point.

BP 2010

In April 2010, the Deep Water Horizon off-shore oil platform suffered a catastrophic failure which led to the direct deaths of 11 workers and the largest marine oil spill in history. Several lessons on ESG data and the decision-making that ultimately led to the failure of Deep Water Horizon are evident from the ensuing investigations.[13]

First, the ESG data produced by BP prior to the incident was not predictive of the underlying or root causes of the accident. BP produced extensive ESG data in its Sustainability Review leading up to the Deep Water Horizon incident.[14] The BP Sustainability Review included data on environmental and safety performance, descriptions of the BP Operating Management System which included controls on safety and environmental performance, policy statements in both areas and an attestation of data accuracy and the validity of the group-level materiality process undertaken by BP issued by Ernst & Young LLP. In terms of indicators, the BP Sustainability Review included indices tracking the Global Reporting Initiative and IPIECA (originally the International Petroleum Industry Environmental Conservation Association)[15] Guidelines for Voluntary Reporting. While these guidelines have evolved since 2010, they remain largely similar in terms of the types of indicators listed. While some have concluded that the Sustainability Review and third-party attestation provide useful information to stakeholders in general and to investigators in hindsight to the Deep Water Horizon disaster, the investigations do not form conclusions on the predictive power of data presented in the BP Sustainability Review other than broad risk pathways identified by the third-party evaluation.[16]

Second, there were a small number of critical decision points leading to the accident. Investigations into the cause of the accident point to a lack of clear data at these decision points that forced decision-makers to weigh caution against cost-savings and convenience. In each decision, the predominant mind-set (caution vs cost) was not explicit to the individuals empowered to make the decision.[17] For example, the authors looked at three fateful engineering decisions where the cautious approach would have resulted in additional costs, additional time, and inconvenient logistics. One of these entailed modeling results from Halliburton that suggested the need for additional centralizers to control gas leakage, which can degrade the cement seal when closing wells. Acquiring additional centralizers would have entailed delays to source and install the proper equipment. This decision was balanced against the perception of uncertainty from the model employed by Halliburton.

In the final analysis of the Deep Water Horizon accident, the core causal factors centered around a lack of clear priorities at key points of decision-making (i.e., making caution for the protection of safety and environment clear priorities over cost efficiency) and the alignment of risk information with decision-making authority (i.e., the person/people making decisions having a full understanding of risk information). At the time of the accident, the information provided to investors and analysts in the Sustainability Review, which was predominantly statements of policy and backward-looking, operational performance data, was not decision-useful to these causal factors.

In the context of ESG data, the BP example points to the fact that past performance as measured by ESG data was not decision-useful for investors or the company. Rather, data on the management control systems and governance structures (G^{ES}) underlying these ESG risks would have presented a more insightful assessment of the risk.

Pacific Gas and Electric 2019

In early 2019, Pacific Gas and Electric (PG&E) filed for bankruptcy in the face of potential civil liabilities due to multiple wildfires in California thought to be started by downed power lines. The Wall Street Journal described the case as the first climate change bankruptcy,[18] as climate change has played a role in increasingly dry conditions recorded throughout the state. However, as the bankruptcy proceedings play out, it has become clear that PG&E recognized the risks of wildfire, and likely the role of climate change in exacerbating that risk,[19] yet faced challenges in addressing the risk appropriately.

In 2010, PG&E was court-ordered to undertake extensive safety investments in the wake of a gas pipe explosion in San Bruno, California. The gas pipeline safety measures resulted, at least partially, in reduced dividends to investors in the subsequent six years and a stock price stagnation.[20] The pressure from investors to focus on return and share price performance has grown for almost a decade.

It is not surprising then that when the same judge from the San Bruno case overseeing PG&E's probation called on the utility to reduce wildfire risk, PG&E responded by claiming that the wildfire risk reduction efforts would be exceedingly expensive ($150 billion) and instead committed to only $2.3 billion in safety measures in 2019. The commitment, documented in the PG&E Wildfire Safety Plan, includes greater equipment inspection rates, tree trimming, and potential black-outs to customers during high risk events.

PG&E's, and indeed its investors', awareness of the increasing risk and its decisions through 2018 echo those of BP leading up to Macondo: competing priorities at the point of decision. In this case, safety investments and a full understanding of the potential magnitude and probability of fire risks that was accelerating under climate change were sacrificed in the face of shareholder pressure for financial returns.

Theoretically, PG&E could have identified fire risk from increased drought frequency in California as a potential climate change risk under the Task Force on Climate-related Financial Disclosure (TCFD) guidance described earlier. It would then have disclosed its efforts to manage the risks through better management practices, higher safety standards, improved monitoring practices, shut-down protocols, insurance coverage, etc. While there is no guarantee that these mitigation efforts would have prevented the risk, the TCFD guidance aims only to increase the amount of decision-useful information available to investors.

As with the BP example, ESG data, for example, carbon emissions data to mitigate climate change risks or health and safety data targeted at mitigating risks from catastrophic gas explosions missed the large potential risk. Management system information (G^{ES}) describing asset maintenance would have been more decision-useful to understand the ability of PG&E to mitigate raising wildfire risk in the face of climate change.

Volkswagen 2015

In 2015, Volkswagen engineered emissions testing cheating software in their popular lines of "clean-diesel" engine passenger cars around the world. The fallout of the scandal led to the resignation and indictment of CEO Martin Winterkorn, jailing of former U.S. Executive Oliver Schmidt, 25% reduction in U.S. auto sales, 37% drop in shares, over $25 billion in fines and long-term reputational damage to the world's and Europe's largest automaker. The cause of the scandal quickly focused on engineers within the company that faced competing incentives to deliver both performance and low emissions.

However, these competing priorities extended throughout the company. Volkswagen had successfully aligned its efforts to lobby European governments with its marketing strategy for European customers around the idea of clean diesel as a solution to both air pollution concerns, as well as customer demands for high performance vehicles. The company was aggressively selling this idea into new markets such as the United States. But in the face of different policy priorities from governments globally, Volkswagen found itself unable to successfully lobby for consistent emissions standards and was therefore facing the fundamental technology trade-off of emissions and performance in internal combustion engines. These competing priorities cascaded throughout the company, from executive to engineer, and were exacerbated by aggressive growth and diversity of global regulations.[21]

When these competing priorities are viewed in the context of the lack of controls and oversight and a corporate culture that did not tolerate excuses, the result is somewhat less surprising. Ferdinand Piëch, the former CEO famously told a journalist asking how he would respond to engineers that could not deliver on promised technology advancements: "Then I will tell them they are all fired and I will bring in a new team. And if they tell me they can't do it, I will fire them, too."[22] Piëch handpicked Winterkorn in 2007 to become the new CEO while remaining as the Board Chair. Between them, the corporate culture focused on success at any costs and strong disincentives to question directions from leadership and managers.

The "right" and "wrong" decisions in the Volkswagen case appear to be more clear-cut than in the PG&E and BP examples suggesting that information on potential risks is potentially less relevant. However, the clarity of prioritization in decision-making was actually working against the engineers in the case of Volkswagen—employees understood the "right thing to do" but faced clear priorities from the corporate culture that this was not the correct decision.

Comparison

In all three examples, the companies in question were widely lauded as the most responsible and sustainable companies in their sectors. Volkswagen, for example, was identified as the most sustainable auto manufacturer according to the Dow Jones Sustainability Index in 2015, the year of the scandal.[23] BP was listed as the second most sustainable large company in the world by the AccountAbility Rating in 2006, down from first in 2005—only four years before the Deep Water Horizon accident and a year after the Texas City refinery accident. PG&E lists a number of recognitions in sustainability from 2016 including the #1 most sustainable electric and gas provider from Newsweek Green Rankings and the #1 utility in Corporate Responsibility Magazine's 100 Best Corporate Citizens ranking. Each company reported (and continues to report) to investors and other stakeholders an impressive array of policies, commitments, management systems, controls, and metrics in sustainability and corporate responsibility.

What are we to make of these examples where companies have been impacted by ESG issues while simultaneously producing world-class disclosures of responsible performance? The lesson for investors is that the ESG metrics collected today are by and large insufficient to determine an accurate picture of the risk faced by companies in the face of rising ESG challenges. Today's ESG metrics are primarily backward-looking performance data. They assess the current status and hygiene of the company rather than the processes undertaken by the company to ensure that performance. More importantly, they reflect performance only against known risks, but not the processes in place to identify and manage emerging risks. The reality is that most ESG risks are currently rising[24] and less well understood than traditional financial risks.[25] As severe weather, water resource constraints, migration, inequality, and other factors rise, how can investors best identify the companies that are most resilient? More specifically, how can investors identify companies that are better able to identify and prevent acute mis-steps in the face of these rising challenges?

The second lesson in these examples is that each highlights a failure of management or governance, specifically the failure on the part of the company to manage or govern an ESG risk. For BP, the failure was on the management of safety as a priority for the company. For PG&E, the failure was on the management of a dynamic risk brought about by climate change. For Volkswagen, the failure was on the management of technology constraints in the face of commitments to address air pollution and energy transitions.

A Proposed Evolution for ESG Metrics

The takeaway for the investor seeking to understand the ESG risks embedded in their portfolio is that *focus needs to be placed on the quality of management practices specific to the most material ESG risks facing the company.* New sets of indicators, associated metrics, and data are needed to assess the ability of companies to appropriately identify ESG risks and the critical points of decision-making for those risks, reveal the quality of the management systems, and demonstrate clarity of priorities at those critical decision-making points.

Two sets of indicators that balance feasibility of collecting information with relevance to investors are needed. The first set of indicators might focus on the quality of the processes for identifying ESG risks. Previous studies have demonstrated that the process of risk assessment and identification in sustainability is more valuable than the disclosed outcomes as companies have strong incentives to over-disclose material issues in the face of wide stakeholder expectations.[26]

Recommended Indicators for ESG Risk Assessment Processes

- What degree of overlap between environmental and social (ES) issues in the Financial Report and Sustainability Report is present?
- What material ES issues are identified in Section 1(a) of the 10-k form or equivalent?
- Does the company use boiler-plate language for ES risks in Section 1(a) of 10-k or equivalent?
- Are ES risks tested in enterprise risk management process?
- Are enterprise risk management criteria aligned to/allow for externalized ES risks?

The second set of indicators attempts to illuminate the breadth and depth of management controls and governance systems that would establish clear priorities for decisions made throughout the company. These indicators include top-level, executive leadership through to procedures to establish, and consider, external stakeholder perceptions.

Recommended Control and Governance Indicators

- Is there a dedicated Board Committee for ES issues?
- Is there an Independent Director with named responsibility for ES issues?
- Is there a Board or Committee Charter with material ES aspects identified?
- What is the gender and cultural diversity of Board?
- Do Board members have ES experience or training?
- Does the company issue a Statement of Significant Audiences and Materiality?
- Is the company a registered Benefit Corporation?
- Is there a Board Charter recognizing interests of multiple stakeholders and/or multiple capital valuations?
- What is the level of employee awareness of group-level responsibilities and priorities for ES issues?
- Does the company assess and report the level of interpersonal trust between employees and line managers?
- Does the company have bypass mechanisms such as stop-work authority, ombudsman, or third-party ethics notifications?
- Does the company have one or more key strategic decision-making processes (e.g., ERM, LCA) with identified mechanisms to integrate external stakeholder information?
- Does executive compensation include material ES aspects?

- Is executive compensation aligned to long-term corporate resilience?
- Are material ES risks included in the operating management systems?
- Are material ES risks included in annual internal audit schedule?
- Is external stakeholder feedback documented and integrated into operating management systems?

Conclusion

There are a number of remaining hurdles that will need to be addressed to operationalize the proposed indicators. Many of these have been previously identified and include questions of comparability, data availability, and data accuracy.[27] Nevertheless, much of the data to populate the suggested metrics are available in audited financial reports (such as Board composition, protections, and experience) or other voluntary ESG reports (such as materiality and risk assessment processes as well as operating management system descriptions). Details of stakeholder engagement, internal audit calendars, and enterprise risk management criteria are likely to prove more difficult to obtain purely through public disclosures.

The proposed indicators are intended to look beyond current performance metrics and backward-looking ESG performance data toward the quality of the underlying risk identification and control mechanisms. They are intended to provide more decision-useful information for investors trying to assess whether companies are exposed to current and emerging environmental and social risks and whether those companies have sufficient control and clear priorities for decision-making processes to mitigate acute risks. However, these proposed indicators are not meant to provide a standard of due diligence. Individual priorities and insight of the investor will continue to determine the effectiveness of the indicators in predicting risk exposure. A careful assessment of these indicators may have provided insights into the risks within BP, Volkswagen, and PG&E, but only through careful analysis and the presence of meaningful data.

Notes

1. Esty, D., & Cort, T. (2017). Corporate Sustainability Metrics: What Investors Need and Don't Get. *The Journal of Environmental Investing*, 8(1); GSIA. (2016). *Global Sustainable Investment Review 2016*. Accessed March 28, 2017, http://www.ussif.org/files/Publications/GSIA_Review2016.pdf.UN; PRI. (2018). *Fiduciary Duty in the 21st Century*. https://www.unpri.org/download?

ac=1378Desclee; Dynkin, A. L., Hyman, J., & Polbennikov S. (2016, October 30). *Sustainable Investing and Bond Returns*. Barclays Research. https://www.inv estmentbank.barclays.com/our-insights/esg-sustainable-investing-and-bond-ret urns.html?trid=39638125&cid=disp_sc00e00v00m04USpa10pv34#tab3; CFA Institute. (2017). *Environmental, Social and Governance (ESG) Survey*. www.cfa institute.org/en/research/survey-reports/esg-survey-2017.

2. Alshehhi, A., Nobanee, H., & Khare, N. (2018). The Impact of Sustainability Practices on Corporate Financial Performance: Literature Trends and Future Research Potential. *Sustainability, 10*(2), 494; Ameer, R., & Othman, R. (2012). Sustainability Practices and Corporate Financial Performance: A Study Based on the Top Global Corporations. *Journal of Business Ethics, 108*(1), 61–79; Borgers, A., Derwall, J., Koedijk, K., & Ter Horst, J. (2013). Stakeholder Relations and Stock Returns: On Errors in Investors' Expectations and Learning. *Journal of Empirical Finance, 22*, 159–175; Cai, L., & He, C. (2014). Corporate Environmental Responsibility and Equity Prices. *Journal of Business Ethics, 125*(4), 617–635; Dhaliwal, D. S., Li, O. Z., Tsang, A., & Yang, Y. G. (2011). Voluntary Nonfinancial Disclosure and the Cost of Equity Capital: The Initiation of Corporate Social Responsibility Reporting. *The Accounting Review, 86*(1); Endrikat, J. (2015). Market Reactions to Corporate Environmental Performance Related Events: A Meta-Analytic Consolidation of the Empirical Evidence. *Journal of Business Ethics*, 1–14; Friede, G., Busch, T., & Bassen, A. (2015). ESG and Financial Performance: Aggregated Evidence from More Than 2000 Empirical Studies. *Journal of Sustainable Finance & Investment, 5*(4), 210–233; King, A. A., & Lenox, M. J. (2001). Does It Really Pay to Be Green? An Empirical Study of Firm Environmental and Financial Performance. *Journal of Industrial Ecology, 5*(1), 105–116; Kitzmueller, M., & Shimshack, J. (2012). Economic Perspectives on Corporate Social Responsibility. *Journal of Economic Literature, 50*(1), 51–84; Klassen, R. D., & McLaughlin, C. P. (1996). The Impact of Environmental Management on Firm Performance. *Management Science, 42*(8), 1199–1214; Kurapatskie, B., & Darnall, N. (2013). Which Corporate Sustainability Activities Are Associated with Greater Financial Payoffs? *Business Strategy and the Environment, 22*(1), 49–61; Rezaee, Z., & Tuo, L. (2017). Voluntary Disclosure of Non-Financial Information and Its Association with Sustainability Performance. *Advances in Accounting, 39*, 47–59; Trudel, R., & Cotte, J. (2009). Does It Pay to Be Good? *MIT Sloan Management Review, 50*(2), 61; Yates-Smith, C. (2013). Socially Responsible Investment: Good Corporate Citizenship or Hidden Portfolio Risk? *Law and Financial Markets Review, 7*(2), 112–117.

3. Eccles, R. G., & Krzus, M. P. (2017, December 14) *An Analysis of Oil & Gas Company Disclosures from the Perspective of the Task Force on Climate-Related Financial Disclosures*. Available at SSRN: https://ssrn.com/abstract=3091232.

4. Khan, M., Serafeim, G., & Yoon, A. (2015, March). *Corporate Sustainability: First Evidence on Materiality*. Harvard Business School.

5. Calvert & HBS. (2018). *How Sustainability Disclosure Is Helping Drive Stock Prices.* https://www.calvert.com/includes/loadDocument.php?fn=26832. pdf&dt=fundpdfs%27MSCI; Calvert & HBS. (2014). *Executive Summary: Intangible Value Assessment (IVA) Methodology.* https://www.msci.com/resources/factsheets/IVA_Methodology_SUMMARY.pdf.

6. Vogel, D. J. (2005). Opportunities for and Limitations of Corporate Environmentalism. *Environmental Protection and the Social Responsibility of Firms: Perspectives from Law, Economics, and Business,* 197; Salzmann, O., Ionescu-Somers, A., & Steger, U. (2005). The Business Case for Corporate Sustainability: Literature Review and Research Options. *European Management Journal,* 23(1), 27–36.

7. CDSB. (2017, June 2). *Uncharted Waters: How Can Companies Use Financial Accounting Standards to Deliver on the TCFD's Recommendations.* https://www.cdsb.net/news/task-force/692/uncharted-waters-how-can-companies-use-financial-accounting-standards-deliver; Eccles, R. G., & Krzus, M. P. (2017, December 14). *An Analysis of Oil & Gas Company Disclosures from the Perspective of the Task Force on Climate-Related Financial Disclosures.* Available at SSRN: https://ssrn.com/abstract=3091232.WBCSD; Eccles, R. G., & Krzus, M. P. (2017, January 18). *Sustainability and Enterprise Risk Management: The First Step Towards Integration.* http://www.wbcsd.org/Projects/Non-financial-Measurement-and-Valuation/Resources/Sustainability-and-enterprise-risk-management-The-first-step-towards-integration.

8. Yale Initiative on Sustainable Finance and ERM. (2019). *Investors Push the Pace of Climate Risk Financial Disclosures.* https://www.erm.com/globalassets/documents/publications/2018/yalecbe-erm-investors-push-the-pace-on-climate-risk-financial-disclosures.pdf.

9. The Committee of Sponsoring Organizations of the Treadway Commission COSO) is a joint initiative five private sector organizations and is dedicated to providing thought leadership through the development of frameworks and guidance on enterprise risk management, internal control, and fraud deterrence.

10. Herz, R. H., Monterio, B. J., & Thomson, J. C. (2018). *Leveraging the COSO Internal Control—Integrated Framework to Improve Confidence in Sustainability Performance Data.* https://www.google.com/url?sa=t&rct=j&q=&esrc=s&source=web&cd=2&ved=2ahUKEwjt3_7zxaPkAhVlkeAKHZNODQMQFjABegQIAhAC&url=https%3A%2F%2Fwww.imanet.org%2F-%2Fmedia%2F73ec8a64f1b64b7f9460c1e24958cf7d.ashx&usg=AOvVaw22wXzYQ2CsUQq6Uqb5qGJ9.

11. COSO and WBCSD. (2018, October). *Enterprise Risk Management: Applying Enterprise Risk Management to Environmental, Social and Governance Risks.* https://www.wbcsd.org/Programs/Redefining-Value/Business-Decision-Making/Enterprise-Risk-Management/Resources/Applying-Enterprise-Risk-Management-to-Environmental-Social-and-Governance-related-Risks.

12. SASB defines materiality based on the definition adopted by the U.S. Supreme Court as information presenting "a substantial likelihood that the disclosure

of the omitted fact would have been viewed by the reasonable investor as having significantly altered the 'total mix' of information made available" (TSC Industries, Inc. v. Northway, Inc., 426 U.S. 438 [1976]).

13. Bartlit Jr., F. H. (2011, February 17). *Chief Counsel, National Commission on the BP Deepwater Horizon Oil Spill and Offshore Drilling* (Macondo Gulf Oil Disaster Chief Counsel's Report 2011); Baker, J., et al. (2007, January). *The Report of the BP U.S. Refineries Independent Safety Review Panel*; Hopkins, A. (2012). *Disastrous Decisions: The Human and Organizational Causes of the Gulf of Mexico Blowout.* Sydney: CCH Australia; Ingersoll, C., Locke, R. M., & Reavis, C. (2012). *BP and the Deepwater Horizon Disaster.* MIT Sloan School of Management, Case Study; Mobus, J. L. (2012). Corporate Social Responsibility (CSR) Reporting by BP: Revealing or Obscuring Risks? *Journal of Legal, Ethical and Regulatory Issues, 15*(2), 35.

14. BP. (2010). *Sustainability Review 2009.* http://www.bp.com/liveassets/bp_int ernet/globalbp/STAGING/global_assets/e_s_assets/e_s_assets_2009/downlo ads_pdfs/bp_sustainability_review_2009.pdf.

15. Originally the "International Petroleum Industry Environmental Conservation Association".

16. Mobus, J. L. (2012). Corporate Social Responsibility (CSR) Reporting by BP: Revealing or Obscuring Risks? *Journal of Legal, Ethical and Regulatory Issues, 15*(2), 35.

17. Ingersoll, C., Locke, R. M., & Reavis, C. (2012). *BP and the Deepwater Horizon Disaster.* MIT Sloan School of Management, Case Study.

18. Gold, R. (2019, January 18). PG&E: The First Climate-Change Bankruptcy, Probably Not the Last. *The Wall Street Journal.* https://www.wsj.com/articles/ pg-e-wildfires-and-the-first-climate-change-bankruptcy-11547820006.

19. Baker, D. R., & Roston, E. (2019, January 22). After PG&E's Climate-Driven Bankruptcy, Who's Next? *Bloomberg.* https://www.bloomberg.com/news/art icles/2019-01-22/why-a-pg-e-bankruptcy-could-change-climate-calculus-qui cktake.

20. Eavis, P., & Penn, I. (2019, February 13). The Struggle to Control PG&E. *The New York Times.* https://www.nytimes.com/2019/02/13/business/energy-enviro nment/pge-wildfire-bankruptcy-control.html.

21. Ewing, J. (2017). *Faster, Higher, Farther: The Inside Story of the Volkswagen Scandal.* Random House.

22. Ewing, J. (2017). *Faster, Higher, Farther: The Inside Story of the Volkswagen Scandal.* Random House.

23. Treehugger. (2020). *Corporate Responsibility.* https://www.treehugger.com/cor porate-responsibility/volkswagen-given-sustainability-award-dow-jones-sustai nability-index.html.

24. This is supported by annual surveys from the World Economic Forum that show environmental and social issues are increasingly identified by global leaders as the most significant threats to the world economy: World Economic

Forum. (2019). *The Global Risks Report 2019.* https://www.weforum.org/rep orts/the-global-risks-report-2019.

25. Esty, D., & Cort, T. (2017). Corporate Sustainability Metrics: What Investors Need and Don't Get. *The Journal of Environmental Investing, 8*(1).

26. Esty and Cort explore the difficulties inherent to the current set of ESG data collected by investors (Esty, D., & Cort, T. [2017]. Corpo- rate Sustainability Metrics: What Investors Need and Don't Get. *The Journal of Environmental Investing, 8*[1]), while the WBCSD looks at the underpinning processes of materiality that are leading to mismatches in the data and investor expectations (WBCSD. [2018]. *Materiality in Corporate Reporting—A White Paper Focusing on the Food and Agriculture Sector.* https://www.wbcsd.org/Programs/Redefining-Value/Resources/A-White- Paper-focusing-on-the-food-and-agriculture-sector).

27. Esty, D., & Cort, T. (2017). Corporate Sustainability Metrics: What Investors Need and Don't Get. *The Journal of Environmental Investing, 8*(1).

4

Creating Investment-Grade Corporate Sustainability Metrics

Daniel C. Esty

Abstract Rising interest in sustainable investing has led to intensified scrutiny of environmental, social, and governance (ESG) reporting. But this heightened interest has convinced many investors and market analysts that the existing ESG metrics lack clear definitional foundations, methodological consistency, analytic rigor, and reliable verification, thus creating doubts about their true validity, comparability, and capacity to distinguish corporate sustainability leaders from laggards. This chapter surveys the gaps and shortcomings in today's ESG data and calls for a mandatory corporate sustainability reporting structure backed by government standards, review, and enforcement. Such a framework ensure that sustainability-minded investors have access to *investment-grade* ESG metrics that inspire trust and confidence.

Keywords Sustainable investing · ESG reporting · Investment-grade metrics · Data quality · ESG methodologies · Validation · Trust · Investor confidence · Greenwash · Materiality · Data normalization · Mandatory reporting · Data comparability

D. C. Esty (✉)
Yale University, New Haven, CT, USA
e-mail: daniel.esty@yale.edu

© The Author(s) 2020
D. C. Esty and T. Cort (eds.), *Values at Work*,
https://doi.org/10.1007/978-3-030-55613-6_4

Interest in *sustainable* investing has risen rapidly.[1] A growing number of investors now want their portfolios to reflect their values, including their interests in addressing climate change, air and water pollution, diversity in the workplace, labor rights, pay equity, good corporate governance, or other issues. While environmental, social, and governance (ESG) metrics exist (with dozens of data providers competing to sell ESG scores to investors), much of the data available does not inspire confidence among fund managers, investment advisors, or investors themselves. Doubts about the quality, methodological underpinnings, comparability, and integrity of the available corporate sustainability metrics represent a major obstacle to an expanded commitment of capital to sustainability-oriented projects, companies, mutual funds, and other investment vehicles.[2] This chapter analyzes what is missing, why so much of the existing ESG data appears to be incomplete and unreliable, and what should be done to create a structure of "investment-grade" sustainability metrics.[3]

Trust emerges as the core issue holding back an expanded commitment of funds to sustainable investing. Investors do not know what they are getting with the existing ESG metrics. They worry that much of the data available in the marketplace reflects *noise* rather than clear *signals* about which companies are sustainability leaders—or worse yet, corporate "greenwash" where claims made do not stand up to scrutiny.[4]

Conceptual Challenges

Several specific conceptual problems can be identified regarding efforts to develop clear, credible, and actionable ESG metrics. First, the term *sustainability* lacks an agreed-upon definition. Indeed, while some investors want information on corporate *environmental* performance, others are more concerned about the commitment of the companies in their portfolios to society's well-being more broadly and to *social* concerns such as racial justice and structural economic inequality. Yet other investors put *governance* practices at the top of their list of critical issues. Fundamentally, every investor will have an individual and unique view of what sustainability means and thus what performance measures ESG data providers should prioritize.[5] With regard to some issues, moreover, investors disagree as to what constitutes the *sustainable* direction or outcome. Does nuclear power, for example, improve or harm our sustainability efforts? Should stem cell research be counted as a positive or a negative?

Likewise, ESG data providers disagree over how to define sustainability, what issues matter most, and how to construct particular metrics. Thus, when two different data providers offer similar sounding datasets that seem to measure the same dimension of sustainability, the metrics they publish will often not match. Indeed, an analysis of the sustainability scoring of MSCI and Sustainalytics, two of the leading ESG data providers, found a correlation of only 0.2.[6] This result—emblematic of the wide disparities in ESG frameworks, assumptions, methodologies, and data management that exist—makes accurate comparisons difficult and contributes significantly to the conclusion that the metrics being offered to investors are insufficiently rigorous to bank on for investment purposes.

Second, significant controversy exists over the degree to which sustainability leadership translates systematically into marketplace success. A number of studies have purported to find a statistically significant relationship between corporate sustainability (measured in a number of different ways) and financial performance.[7] But other analyses have picked apart these studies and come to the conclusion that a broad-based link between sustainability and marketplace outperformance cannot be established.[8]

What seems likely is that *some aspects* of sustainability leadership show a relationship with financial success, but other aspects of this multi-dimensional concept do not correlate with outperformance. Quite clearly, companies with sophisticated energy management systems, such as Walmart, will use this *eco-advantage* to cut costs and advance their competitive posture in the marketplace.[9] But equally clearly, being out in front on aspects of sustainability that impose costs could diminish a company's prospects for marketplace success in a world of incomplete regulation in which externalities (such as pollution) are not fully internalized,[10] thus rewarding *unsustainable* corporate behavior.[11] This leads to the conclusion, as Professor Dixon-Fowler has noted, that asking "does it pay for companies to be green?" makes little sense. The right inquiry would be: "*when* does it pay to be green?"[12]

Third, these discrepancies have sharpened the focus on which issues—and thus which ESG metrics—are *material* for the purposes of sustainability scoring. The rate of office paper recycling (historically a favorite gauge of many environmental groups) is, for example, unlikely to be truly indicative of anything important. But the emphasis on *materiality* has helped to highlight some additional methodological choices over which ESG commentators, data providers, and investors do not necessarily agree. The Task Force on Climate-related Financial Disclosures (TCFD), launched by the then-Governor of the Bank of England Mark Carney, has advanced, for instance, a definition of materiality that emphasizes company-specific disclosure of risks and opportunities related to climate change in accordance with national reporting requirements, which tends to emphasize the potential for information to have a substantial short-term effect on the price of a company's stock. Likewise, the Sustainable Accounting Standards Board (SASB) has adopted the U.S. Supreme Court's definition of *financial materiality*, which centers on disclosure of information that a "reasonable investor" would view as significant in making an investment decision. This frame of reference has led SASB to identify a small number of material issues by industry (sometimes as few as three categories out of 26 possible sustainability-related business issues). But this definition has led many others to critique the SASB framework as too narrow and short-term in focus. One might ask in this regard whether the SASB reporting structure would have captured the corporate ethics failures of Volkswagen or the human resource management shortcomings of Uber?

Others, such as the Global Reporting Initiative, have argued for a broader and longer-term view of materiality that goes beyond immediate financial impact to consider issues and trends, such as climate change, that might substantially affect a company's posture (defined more broadly than its stock value) over time. This more comprehensive perspective—encompassing a company's growth, profitability, capital efficiency, and risk profile, with reporting requirements structured on an industry-by-industry basis— will almost certainly serve as a better starting point for ESG reporting going forward than materiality definitions that focus on short-term financial impacts alone. Indeed, many of today's sustainability-minded investors care about corporate performance on critical issues, such as climate change, even in circumstances where there will be no likely effect on the company's stock value.

In this spirit, David Lubin and I have argued in the past for ESG reporting that covers the full range of issues that sustainability investors will want to know about,[13] with special emphasis on activities that might affect a company's business value across a core set of factors shaping business success: Growth, Productivity, and Risk.[14] This *value-driver* model looks not just at the down-side risks, but also the upside opportunities for companies providing solutions to sustainability challenges such as climate change.[15]

Fourth, because of the financial frame of most past ESG metrics efforts, a critical dimension of corporate sustainability has tended to be ignored: how well a company delivers on society's goals as well as its own. As the Business Roundtable (a confederation of 200 major U.S. companies) recently acknowledged with its revised "statement of purpose of a corporation," the days when companies could follow Milton Friedman's famous admonition to focus narrowly on optimizing shareholder value have passed.[16] In our era of "shared value"[17] and "conscious capitalism,"[18] companies must pay attention to a broader set of stakeholders and ensure that they deliver on behalf of their customers, employees, suppliers, communities, and society as a whole—as well as delivering profits to their shareholders.

In this regard, a growing number of investors now want a clearer picture of which companies are meeting society's need as well as delivering stock market gains—recognizing that those that fall short of public expectations risk becoming political targets and losing their *social license to operate*.[19] The prospect of getting crosswise with the public and political leadership cannot be dismissed lightly, as Uber found out when it was kicked out of London in 2017. This new interest means that more emphasis will need to be placed on metrics that track corporate contributions to society's goals as framed by structures such as the UN Sustainable Development Goals (SDGs).[20] While not all of the 17 goals and 169 specific targets beneath them are issues that companies might be held responsible for, some dimensions of the SDG structure must now be prioritized within corporate reporting frameworks.

Fifth, because many other factors besides the sustainability strategy of a company determine corporate financial performance, we cannot be sure that ESG leaders will consistently outperform their peers. Market dynamics, success with regard to technology development, production, distribution, marketing, cost controls, customer experience, and other factors may overwhelm sustainability leadership in determining which companies are profitable and gain market share in any particular industry.

But as I have argued (along with my colleague David Lubin), success in managing the *sustainability imperative* that is sweeping across the business world and society more broadly may provide a valuable leading-edge signal of management quality.[21] We have therefore developed a set of refined ESG metrics that measure a company's *maturity* and *momentum* in addressing sustainability challenges, such as climate change.[22] We believe that sustainability leadership in this regard offers a valuable gauge of the capacity of CEOs and their leadership teams to deliver outperformance more generally.

Finally, not all sustainability investors are the same. Some have a narrow focus perhaps centered on climate change or corporate diversity or labor rights. Others have a wider range of values that they want to reflect in their portfolios.[23] Some sustainability-minded investors think of themselves as *impact investors* who care as much about changing society as they do about the financial returns from their investments. More generally, a growing number of sustainability investors consider themselves *value* oriented and expect to earn returns that match traditional benchmarks. Others recognize that by screening companies based on their sustainability performance, they will narrow their investable universe, which may increase the volatility of their portfolios, resulting over time in sub-par returns.

Given this diversity of sustainability priorities, investment goals, and risk tolerances, no single tightly framed structure of ESG metrics will suffice. Indeed, attempts to offer generic sustainability investment vehicles have run into trouble.[24] Instead, the ESG data framework of the future needs to be understood as a menu of sustainability metrics that cumulatively meet the full range of investor needs—recognizing that individuals will rely on the subset of metrics that align with their own interests and investment strategies. But while investors vary in the metrics they want to use, almost all want decision-useful ESG signals that offer methodological rigor, consistency with recognized standards, transparency with regard to underlying assumptions, and high integrity.

ESG Data Shortcomings

Much of the ESG scoring presently available to investors has significant limitations. These shortcomings make it difficult to be confident that the metrics provided truly separate sustainability leaders from laggards. These structural and methodological weaknesses include:

1. *Inconsistent reporting frameworks*—The lack of standardized methodologies and best practices for ESG metrics takes many forms including:

   - divergent time frames and lack of regular data updating, which makes comparisons across companies nearly impossible, especially benchmarking that relies on data from multiple providers that have not been aligned;

   - unclear ESG methodological standards, which means that even broadly understood topics, such as "greenhouse gas emissions," will suffer from wide variations in what companies report as some provide just Scope 1 emissions while others might report Scopes 1 and 2 and the most diligent companies report Scopes 1, 2, and 3;

   - the reality that even within specific categories with clear methodological definitions, significant inconsistencies emerge as companies choose different ways to report—such as Tata Motors falsely appearing to be an auto industry sustainability leader because the company reported only the emissions from its business travel in Scope 3 rather than the emissions of the vehicles it sold;

– failure to distinguish between missing data and poor performance—often recording both as zero;

– incomplete reporting—meaning that as many as 50% of the cells in the data matrices of major ESG data providers (such as MSCI, Bloomberg, and Sustainalytics) are blank;

– inconsistent gap-filling with some data providers leaving gaps empty while others use extrapolation or other modeling techniques to fill holes in their data matrices—but often without making clear the assumptions that go into their gap-filling strategies;

– lack of clarity on whether gaps are filled with some relevant *average* number or whether missing fields are filled with a *penalized* score (perhaps reflecting the 20th percentile of similarly situated companies that are reporting) to create an incentive for future disclosure;

– limited efforts to detect errors and anomalies that arise from poor quality control on information gathering, data entry, mislabeled companies, failure to use correct company "identifiers" to connect data to the right entity, and much more;

– lack of transparency on which metrics derive from *measurements* and actual "on-the-ground" results and which come from *surveys* or data that have been *modeled;*

– misaligned scopes of reporting with some companies providing data on a consolidated basis (including all subsidiaries across all geographies) while others report on a more disaggregated or limited basis.

2. *Failure to normalize data*—Absent concerted efforts to ensure that comparisons are "apples to apples," ESG data providers may seriously distort rankings by failing to guarantee that company scores reflect common methodologies and assumptions including the need to:

– use common units (such as pounds vs. kilos vs. tons);

– standardize comparative analyses to avoid distortions that arise from varying degrees of vertical integration (so as not to penalize companies who own their supply chains and reward those who have outsourced "dirty" aspects of production)—with the goal being reporting that reflects the entire value chain;

– consolidate reporting so that the information published covers all of a company's facilities in all the geographic locations in which it operates as well as all of its subsidiaries—and provide full transparency on the scope of reporting;

- recalibrate reports to reflect mergers, acquisitions, and asset sales that change the baseline from which company progress should be gauged;

- use appropriate denominators to ensure that scores do not simply reflect the size of a company—so that large enterprises, such as GE with four environmental incidents reported and $120 billion in global sales, do not get a worse score than a company with $2 billion in revenues and two incidents—but rather look at the ratio of incidents/dollar of revenue, which would be 1/$30 billion for GE and 1/$1 billion for its smaller competitor;

- ensure that conglomerates are associated with the correct industries for purposes of cross-company benchmarking—with the best practice being to allocate a multi-industry company to each of the sectors it is part of on the basis of sales in that industry.

3. *Lack of Validation and Verification*—Many ESG data providers simply pass along the numbers provided by companies or posted online, with little or no attempt to verify the accuracy of the reporting, rather than seeking:

- third-party auditing and verification of all of the numbers put forward in corporate sustainability reports;

- traceability of each number provided with clarity about the methodological assumptions that underpin the reported figures.

4. *Reliance on static versus dynamic metrics*—The use of metrics that represent a single point in time *snapshot* instead of multi-year data misses the critical *trend* and *momentum* factors in ESG performance such that two companies with the same score might be seen as performing equally well when, in fact, one's performance is deteriorating and the other is improving—meaning that ESG scoring should:

- include trend data, such as five-year average scores, so that ESG data users can see which companies are improving (and how quickly) and which are deteriorating on specific issues;

- ensure that industries as well as companies can be assessed with regard to which are addressing critical social concerns—revealed in improving results—and which are ignoring them.

5. *Focus on backward-looking rather than forward-looking metrics*—The existing ESG data frameworks are heavily slanted toward past results rather than providing a basis for assessing likely future performance—so more forward-looking analytic tools are needed which would:

   – avoid over-reliance on past perceptions and *reputational* scores (such as how many negative stories appeared online or other gauges of stake-holder perceptions from media reports) and put more emphasis on *operational* factors such as advances in eco-efficiency and signals of readiness to respond to sustainability challenges, including movement toward renewable energy sourcing;

   – adopt a *value-driver* framework that tracks how a company's sustain-ability strategy contributes to its growth (both top- and bottom-line), productivity (cost cutting through energy efficiency, etc.), and risk management (by reducing carbon exposure and other sustainability-related risks).

6. *Over-emphasis on downside exposure and not upside opportunities*—Current ESG metrics frameworks focus heavily on risks, pollution levels, and other *negative* impacts—and do far too little to gauge corporate positioning as a sustainability solutions provider, so more metrics are needed that:

   – go beyond carbon or ecological *footprints* to measure corporate *hand-prints*, which track a company's success in delivering sustainability-derived value to its customers—building on the reporting of companies such as Alcoa (now Arconic), which estimates the energy savings that its customers, such as Boeing and Ford, have achieved using its advanced materials;

   – track a company's sustainability-driven sales and whether the growth and profitability of this set of goods and services is faster or slower than other lines of business.

Fundamentally, ESG metrics need to be made more clear, consistent, and reliable—as only a reporting framework that meets investor expectations in this regard will enable a broader commitment to sustainable investing.

Path Toward Investment-Grade ESG Metrics

As noted above, what sustainability-minded investors generally want is a menu of consistently defined and rigorously gathered ESG metrics—all of which provide a reliable foundation for cross-company comparisons and thus portfolio choices.[25] Because disclosure requirements impose a data-gathering burden on companies and thus a cost on society, any ESG reporting framework needs to center on a thoughtfully designed set of mandatory reporting elements tied to *material* issues, while still providing a mechanism for those companies who wish to report more broadly to be able to do so. Likewise, because the list of materiality sustainability issues will vary somewhat from sector to sector, the framework will need to reflect a core set of metrics on which all companies report as well as some industry-specific requirements.

Movement toward a framework of investment-grade sustainability metrics might therefore include three elements: (1) an ESG reporting framework that specifies a set of metrics on which reporting is mandatory, (2) compulsory methodological standards for each issue on which disclosure is mandated, and (3) a trusted validation system.

ESG Reporting Framework

The reporting framework might best be understood as having three tiers of ESG metrics.[26] The *first tier* would specify a core set of mandatory disclosure elements covering environmental and social indicators that have the prospect of being material from *either* a financial or societal perspective over the next decade. This framework would thus encompass environmental issues such as greenhouse gas emissions, waste management, air pollution, water quality and quantity, chemical safety, and land use as well as social concerns such as diversity in both management and the workforce more broadly, investment in human capital, product and production health and safety, community and customer relations, privacy protection, and corporate integrity.

The prescribed time frame would go beyond current reporting requirements by requiring disclosure on issues that might become significant over the intermediate term (out 5–10 years), resulting in a core ESG framework with about 15–20 mandatory items. It would also be widened by including metrics designed to gauge corporate success in delivering sustainability solutions (such as the revenue derived from sales of goods and services that address a customer's energy or environmental challenges).

The *second tier* of reporting would be industry-specific and include metrics that are relevant in narrow circumstances but need not be applied to all companies in all sectors. Thus, for example, in the transportation sector (airlines, freight hauling, and vehicles), it would make sense to track company efforts to shift away from fossil fuels. Likewise, in the mining industry, reporting on overburden management would be critical, but the issue is not relevant in other sectors. This second tier might also provide a framework for company-specific voluntary reporting through which corporate leaders could highlight elements of their sustainability strategies, which they believe set themselves apart from the pack.

The *third tier* would specify a set of governance issues—going above and beyond the sustainability elements of tiers one and two—on which disclosure would be mandatory. This list would cover a limited set of critical institutional issues that gauge a company's capacity for internal oversight and good governance including the structure of its board of directors, the separation of the roles of CEO and Board Chair, the presence of independent directors, as well as the breadth and depth of corporate transparency and reporting.

ESG Methodological Standards

As noted above, the foundation for expanded sustainable investing would be greater confidence in the integrity, comparability, and analytic rigor of ESG metrics. Building such a base of confidence requires not just a common reporting framework but also methodological standards that provide assurance that the numbers reported by Company A can be compared to those of Company B. In the realm of corporate accounting this sort of assurance is provided in the United States by the Financial Accounting Standards Board (FASB), which defines accounting rules through the publication of the Generally Accepted Accounting Principles (GAAP), and which publicly traded companies are required to follow. The standards are reinforced by the norms and practices of the major accounting firms, who recognize that their reputations and continued business viability depend on upholding the standards and verifying corporate compliance punctiliously.[27] Something similar will be required in the sustainability domain, perhaps building on the work Sustainability Accounting Standards Board (SASB) and Global Reporting Initiative (GRI) have done—but with an expanded focus on addressing the full array of problems outlined above.

Validation System

In addition to clear methodological rules for ESG metrics, there needs to be a validation system in place to verify the data that companies report. With regard to financial accounting, the requirement that financial reports be audited and verified by certified accountants provides this structure—with the background threat that inaccurate data or reports could be called to task by legal action from the Securities and Exchange Commission (SEC) or through shareholder suits based on SEC filings. Given this existing SEC role, the easiest path toward ESG validation would be to make the mandatory ESG framework a part of standard corporate financial reporting—requiring certification by auditors, with the threat of legal action under SEC rules providing an added incentive for careful compliance.

Voluntary reporting frameworks could be organized by investment industry groups or perhaps the Sustainability Stock Exchanges Initiative— and could provide an alternative pathway to standardized reporting and a defined set of methodological underpinnings. But such voluntary efforts might fall short of the goal of full investor confidence without some backing by a government entity that can provide a threat of legal penalties where companies violate the requirements. Indeed, investor-oriented corporate data and reports, including ESG metrics, are widely recognized as public goods, which tend to be underprovided absent efforts by government to address the need and over-come the likely market failure that otherwise occurs.[28]

Conclusion

The need to move the data and information available to sustainability-minded investors from the patchwork that exists today toward a structure of investment-grade ESG metrics looms large if the interest in sustainable investing is to be maintained and expanded. The path toward corporate ESG reporting that offers investors a framework of clearly specified, methodologically consistent, and carefully verified signals of sustainability leadership will be bumpy. But success will benefit not just the growing number of sustainable investors who want better alignment between their values and their portfolios, but also society as a whole, as trusted signals of ESG leadership would channel capital toward enterprises that are moving the world toward a more sustainable future.[29]

Notes

1. Esty, D. C., & Cort, T. (2017). Corporate Sustainability Metrics: What Investors Need and Don't Get. *Journal on Environmental Investing*, 8(1), 11–53; Eccles, R., & Klimenko, S. (2019, May–June). The Investor Revolution. *Harvard Business Review*, 106–116.
2. Lewis, E., et al. (2016). Navigating the Sustainable Investment Landscape. *World Resources Institute*, 21; Krosinsky, C. (2017). The State of ESG Data

and Metrics: The Editor's Word. *Journal on Environmental Investing, 8*(1), 5–6; Bose, S., & Springsteel, A. (2017). The Value and Current Limitations of ESG Data for the Security Selector. *Journal on Environmental Investing, 8*(1), 54–73; Esty, D. C., & Karpilow, Q. (2019). Harnessing Investor Interest in Sustainability: The Next Frontier in Environmental Information Regulation. *Yale Journal on Regulation, 36*(2), 655.

3. By "investment grade" ESG metrics, I am referring to a level of data quality and methodological rigor that would permit company-to-company comparisons that instill confidence and trust in investors such that they would be willing to commit funds based on sustainability factors.

4. Dahl, R. (2010). Green Washing: Do You Know What You're Buying? *Environmental Health Perspectives, 118*(6), A246–52. https://www.ncbi.nlm.nih.gov/pmc/articles/PMC2898878/; Esty, D. C., & Winston, A. (2009). *Green to Gold: How Smart Companies Use Environmental Strategy to Innovate, Create Value, and Build Competitive Advantage* (p. 248). Wiley.

5. Esty, D. C., & Cort, T. Corporate Sustainability Metrics: What Investors Need and Don't Get, 13; Vittorio, A. (2016, July 26). Investors to SEC: We Want Better Sustainability Reporting. *Bloomberg BNA.* https://www.bna.com/invest ors-secwe-better-n73014445306 [https://perma.cc/FE2N-74GD].

6. Sindreu, J., & Kent, S. (2018, September 21). Why It's So Hard to Be an 'Ethical' Investor. *Wall Street Journal.*

7. Orlitzky, M., Schmidt, F., & Rynes, S. (2003). Corporate Social and Financial Performance: A Meta-Analysis. *Organization Studies, 24*(3), 403–441; Eccles, R. G. Ioannou, I., & Serafeim, G. (2014). The Impact of Corporate Sustainability on Organizational Processes and Performance. *Management Science, 60*(11), 2835–2857; Friede, G., Busch, T., & Bassen, A. (2015). ESG and Financial Performance: Aggregated Evidence from More Than 2000 Empirical Studies. *Journal of Sustainable Finance & Investment,* 5(4), 210–233; Borgers, A., et al., (2013). Stakeholder Relations and Stock Returns: On Errors in Investors' Expectations and Learning. *Journal of Empirical Finance, 22,* 159–175; Dimson, E., et al. (2015). Active Ownership. *Review of Financial Studies, 28*(12), 3225–3268; Shahzad, A., & Sharfman, M. P. (2015). Corporate Social Performance and Financial Performance: Sample-Selection Issues. *Business and Society, 56*(6); Gartenberg, C., Prat, A., & Serafeim, G. (2016, September). *Corporate Purpose and Financial Performance* (Harvard Business School Working Paper, No. 17-023). Ueng, C. J. (2015). The Analysis of Corporate Governance Policy and Corporate Financial Performance. *Journal of Economics and Finance,* 1–10; Clark, G. L. (2015). *From the Stockholder to the Stakeholder: How Sustainability Can Drive Financial Outperformance.* University of Oxford; Ioannou, I. (2019). *Corporate Sustainability: A Strategy?* (Harvard Business School Accounting and Management Unit Working Paper No. 19-065). https://www.hbs.edu/faculty/Publication%20Files/19-065_16deb9d6-4461-4d2f-8bbe-2c74b5beffb8.pdf.

8. Vogel, D. J. (2005). Is There a Market for Virtue?: The Business Case for Corporate Social Responsibility. *California Management Review, 47*(4), 19–45; Marcus, A. A. *Innovations in Sustainability: Fuel and Food.* Cambridge: Cambridge University Press, 2015; Flammer, C. (2015). Does Corporate Social Responsibility Lead to Superior Financial Outcomes? A Regression Discontinuity Approach. *Management Science, 61*(11), 4–6; Kim, K.-H. Kim, M., & Qian, C. (2018, March). Effects of Corporate Social Responsibility on Corporate Financial Performance: A Competitive-Action Perspective. *Journal of Management, 44*(3), 1097–1118; Kitzmueller, M., & Shimshack, J. (2012). Economic Perspectives on Corporate Social Responsibility. *Journal of Economic Literature, 50*(1), 51–84; Kotchen, M. J., & Moon, J. J. (2012). Corporate Social Responsibility for Irresponsibility. *The B.E. Journal of Economic Analysis & Policy, 12*(1), Article 55; Orlitzky, M. (2013). Corporate Social Responsibility, Noise, and Stock Market Volatility. *Academy of Management Perspectives, 27*(3), 238–254; Eccles, R. G., Ioannou, I., & Serafeim, G. (2014). The Impact of Corporate Sustainability on Organizational Processes and Performance. *Management Science, 60*(11), 2835–2857; Lourenco, et al. (2014). The Value Relevance of Reputation for Sustainability Leadership. *Journal of Business Ethics, 119*(1), 17–28; Kurapatskie, B., & Darnall, N. (2013). Which Corporate Sustainability Activities Are Associated with Greater Financial Payoffs? *Business Strategy and the Environment, 11*(1), 49–61; Ameer, R., et al. (2012). Sustainability Practices and Corporate Financial Performance: A Study Based on the Top Global Corporations. *Journal on Business Ethics, 108*, 61–79.

9. Esty, D. C., & Simmons, P. J. (2011). *The Green to Gold Business Playbook: How to Implement Sustainability Practices for Bottom-Line Results in Every Business Function.* Wiley.

10. King, A., & Lenox, M. (2002). Exploring the Locus of Profitable Pollution Reduction. *Management Science, 48*(2), 289–299; Albertini, E. (2013). Does Environmental Management Improve Financial Performance? A Meta-Analytical Review. *Organization and Environment, 26*(4), 431–457; Endrikat, J. (2015). Market Reactions to Corporate Environmental Performance Related Events: A Meta-Analytic Consolidation of the Empirical Evidence. *Journal of Business Ethics*; Kim, Y., & Statman, M. (2012). Do Corporations Invest Enough in Environmental Responsibility? *Journal of Business Ethics, 105*(1), 115–129; Cai, L., & He, C. (2014). Corporate Environmental Responsibility and Equity Prices. *Journal of Business Ethics, 125*(4), 617–635; Cort, T., & Esty, D. C. (2020). ESG Standards: Looming Challenges and Pathways Forward. *Journal of Organization and Environment.*

11. I have argued elsewhere (Esty, D. C. 2017). Red Lights to Green Lights: From 20th Century Environmental Regulation to 21st Century Sustainability. *Environmental Law, 47*[1], 24) that the "end of externalities" is coming and that business models which depend on externalizing costs onto society—whether in the form of pollution or inadequate wages that leave workers dependent on social safety nets—will be ever more difficult to sustain in the years ahead

as regulators sharpen their policy tools and seek to ensure that companies bear fully the costs that they might otherwise impose on society at large. But how fast this transformation unfolds remains in question and therefore sub-par sustainability practices might be rewarded in stock markets for some period of time to come.

12. Dixon-Fowler, H. R., et al. (2013). Beyond 'Does it Pay to be Green?' A Meta-Analysis of Moderators of the CEP-CFP Relationship. *Journal of Business Ethics, 112*(2), 354.

13. Cort, T., & Esty, D. C. ESG Standards: Looming Challenges and Pathways Forward.

14. Lubin, D. A., & Esty, D. C. (2014, June 17). Bridging the Sustainability Gap. *MIT Sloan Management Review.* https://sloanreview.mit.edu/article/bridging-the-sustainability-gap/.

15. Lubin, D., & Krosinsky, C. The Value Driver Model: A Tool for Communicating the Business Value of Sustainability. *Principles for Responsible Investing* and *UN Global Compact (2013).*

16. Business Round Table. *Statement on the Purpose of a Corporation.* https://opportunity.businessroundtable.org/wp-content/uploads/2019/09/BRT-Statement-on-the-Purpose-of-a-Corporation-with-Signatures.pdf.

17. Porter, M. E., & Kramer, M. R. (2011, January). Creating Shared Value. *Harvard Business Review.* https://hbr.org/2011/01/the-big-idea-creating-shared-value.

18. Mackey, J., & Sisodia, R. (2013). *Conscious Capitalism: Liberating the Heroic Spirit of Business.* Brighton, MA: Harvard Business Review Press.

19. Gunningham, N., et al. (2004). Social License and Environmental Protection: Why Businesses Go Beyond Compliance. *Law and Social Inquiry, 29*(2), 307–341; Global, E. (2017, March 9). *Why Sustainable Development Goals Should Be in Your Business Plan.* https://www.ey.com/en_gl/assurance/why-sustainable-development-goals-should-be-in-your-business-plan.

20. Rosati, F., & Diniz Faria, L. G. (2019). Business Contributions to the Sustainable Development Agenda: Organizational Factors Related to Early Adoption of SDG Reporting. *Corporate Social Responsibility and Environmental Management, 26*(3).

21. Lubin, D. A., & Esty, D. C. (2010, May). The Sustainability Imperative. *Harvard Business Review.* https://hbr.org/2010/05/the-sustainability-imperative.

22. See Constellation Research and Technology (www.constellationresearch.com): For more detail on this M2 (momentum and maturity) model, see Lubin, Moorhead, and Nixon G250 Report. Lubin, D., Moorhead, J., & Nixon, T. (2017). Global 250 Greenhouse Gas Emitters: A New Business Logic. *Thomson Reuters.* https://blogs.thomsonreuters.com/sustainability/wp-content/uploads/sites/15/2017/10/GLOBAL-250-GREENHOUSE-GAS-EMITTERS-A-New-Business-Logic.pdf.

23. Esty and Cort (2017) presented a typology of sustainability investors across several factors.

24. Pinchot, A. C., McClamrock, J., & Christianson, J. (2018, February). *Blackrock CEO Wants Companies To Be Socially Responsible. Here are 6 Ways He Can Show He's Serious.* World Resources Institute. https://www.wri.org/blog/2018/02/blackrock-ceo-wants-companies-be-socially-responsible-here-are-6-ways-he-can-show-he-s. Pinchot, A. C., McClamrock, J., & Christianson, J. (2019, April). Is BlackRock Serious About Sustainability? *GreenBiz.* https://www.greenbiz.com/article/blackrock-serious-about-sustainability.

25. Some investors do not mind the messiness of the existing ESG reporting on the theory they can make better sense of the mess than their competitors.

26. For a deeper discussion of the three tiers of ESG reporting, see Esty et al. (2020, September). *Toward Enhanced Sustainability Disclosure: Identifying Obstacles to Broader and more Actionable ESG Reporting.* Yale Initiative on Sustainable Finance White Paper.

27. Auditing firms that fall short of this standard fall by the wayside as happened to Arthur Andersen in the wake of the Enron accounting scandal. Rauterkus, S. Y., & Kyojik "Roy" Song. (2005). Auditor's Reputation and Equity Offerings: The Case of Arthur Andersen. *Financial Management, 34*(4) (Winter), 121–135. Knechel, W. R. (2013). Do Auditing Standards Matter? *Current Issues in Auditing, 7*(2), A1–A16.

28. Williams, C. A., & Fisch, J. E. *Petition for Rulemaking on Environmental Social and Governance Disclosure.* https://www.sec.gov/rules/petitions/2018/petn4-730.pdf; Esty, D. C., & Karpilow, Q. (2019). Harnessing Investor Interest in Sustainability: The Next Frontier in Environmental Information Regulation. *Yale Journal on Regulation, 36*(2), 625–693; International Organization of Securities Commissions. (2019, January 18). *Statement on Disclosure of ESG Matters by Issuers.* https://www.iosco.org/library/pubdocs/pdf/IOSCOPD619.pdf.

29. Esty, D. C., & Karpilow, Q. Harnessing Investor Interest in Sustainability: The Next Frontier in Environmental Information Regulation, at 686.

5

Asset-Level Physical Climate Risk Disclosure

Natalie Ambrosio Preudhomme and Emilie Mazzacurati

Abstract As businesses increasingly incur the costs of extreme weather events and chronic stresses driven by climate change, there is a need for transparency around corporate exposure to physical climate risks. Understanding risk exposure is an important first step toward managing risk and building corporate resilience. This chapter discusses pharmaceutical companies as a case study to examine how risk exposure aligns with corporate risk disclosures. We compare asset-level climate change risk exposure based on Four Twenty Seven's database of corporate facilities' risk to climate hazards with the risks the companies report in their financial filings and CDP reports. We find a lack of detailed, forward-looking, and consistent risk disclosure by corporations—and that the information disclosed does not align with the risks identified in Four Twenty Seven's separately established climate change risk analysis. The chapter concludes by discussing elements of risk disclosure that companies can improve to provide more decision-useful information to investors, including consistent risk assessment methodologies, risk metrics, and systematic risk management strategies.

Four Twenty Seven is a non-credit rating agency affiliate of Moody's and is a publisher and provider of data, market intelligence and analysis related to physical climate and environmental risks. The views expressed in this chapter are solely those of Four Twenty Seven and do not reflect the opinions of Moody's Investors Service.
© Four Twenty Seven, Inc. Reprinted with permission. All Rights Reserved. The full terms and conditions applicable to this chapter are available at the following link: 427mt.com/copyright.

N. Ambrosio Preudhomme (✉) · E. Mazzacurati
Four Twenty Seven, Inc., Berkeley, CA, USA
e-mail: nambrosio@427mt.com

© The Author(s) 2020
D. C. Esty and T. Cort (eds.), *Values at Work*,
https://doi.org/10.1007/978-3-030-55613-6_5

Keywords Four Twenty Seven · Climate change · Climate risk · Risk exposure · Physical climate risk · Climate change risk · CDP · Extreme weather · Water stress · Pharmaceutical companies · Task Force on Climate-Related Financial Disclosures · Climate risk disclosure

The United States experienced 14 distinct billion-dollar disasters in 2018, making 2018 the fourth most costly year in terms of national disasters. Before the end of summer 2019, high temperature records had been broken multiple times across Europe and in parts of the United States, and July 2019 was the hottest July on record globally.[1] The International Labor Organization estimates that a global temperature rise of just 1.5 degrees Celsius by 2100 would result in lost working hours equal to 80 million full-time jobs, and economic losses of $2.4 trillion by 2030.[2] Extreme weather events such as hurricanes, typhoons, floods, and heat waves, and chronic stresses such as droughts and sea level rise, all threaten corporate assets and business operations with implications for the global economy.

Investors and regulators express a growing concern over climate change impacts on the financial sector and are increasingly calling for more transparency around firms' climate change risks. This transparency is vital so that risk can be accurately priced into investments to avoid a climate-driven financial crisis. Recent events, such as some banks selling mortgages tied to properties exposed to flood risk to public entities, thus shifting risk to taxpayers,[3] and insurance companies' tendency to remove coverage from areas hit by wildfires,[4] suggest that such a financial crisis is a very real possibility. In 2017, the Financial Stability Board's Task Force on Climate-related Financial Disclosures (TCFD) catalyzed the conversation around risk disclosure by providing recommendations for the disclosure of transition and physical risks from climate change.[5]

Transition risks refer to risks that companies face from policies implemented to transition to a low-carbon economy and are most pressing for companies with large dependencies on energy and fossil fuels. For example, as the world shifts away from fossil fuels and oil in particular, an automotive company could see increased production costs due to a carbon price, together with decreasing demand for vehicles if they have not invested well ahead of time in producing electric vehicles. *Physical risks* refer to the physical impacts of climate-driven events, such as floods and heat waves, on businesses through disruptions to their assets' operations, supply chains, and consumer behavior. While significant research and thinking has gone into the development of carbon benchmarks and transition risk, physical climate risks present distinct challenges and they have been the topic of less research to date.

This differentiation is partly because translating climate change projections into economic and financial impacts presents significant challenges due to the sheer volume and format of climate data, the long time frames of climate projections, the varying levels of detail available, and financial users' need for asset-level data that is also globally comparable.[6] While regulations around greenhouse gas emissions will affect entire jurisdictions, an asset's exposure to extreme rainfall or sea level rise is based on its precise geographic location. Future conditions will differ from historic conditions, and the breadth of financial implications is not yet fully known. However, the financial sector is increasingly calling for research and action around understanding these financial implications and preparing global economies for these changes.

Four Twenty Seven, an affiliate of Moody's and a publisher of data on the physical risks of climate change, has developed a dataset that matches the precise locations of corporate assets with granular climate and environmental risk data globally. Analyzing this corporate exposure data, alongside financial filings and corporate risk disclosures, shows that the risks posed by physical climate hazards are not being reported in a clear and comparable format that is meaningful to investors. This chapter uses pharmaceutical and biotechnology companies as a case study to explore current practices and discuss implications for transparency and pricing of physical climate risk into financial markets. We first discuss why climate risk is particularly relevant for the pharmaceutical and biotechnology sector, then highlight examples of ways climate hazards have already affected these companies, and finally discuss whether these companies are thoroughly reporting these risks.

Climate Risk in Pharmaceutical Companies

A Vulnerable Sector

The pharmaceutical sector provides an interesting case study of the many ways in which manufacturing and value chains can be affected by climate-related hazards. Like in most manufacturing sectors, in the pharmaceutical sector, extreme weather can cause costly interruptions in complex value chains and damage high-value equipment, with cascading effects down the value chain when downstream facilities are not able receive the necessary parts or ingredients for their own products or medical operations. In addition, pharmaceutical companies must store their products and ingredients at very precise temperatures and rely on manufacturing facilities with high energy demands and rigid supply chain dependencies, making them vulnerable to disruptions in upstream facilities. By the nature of their products, pharmaceutical companies also play a critical role in public health, and business disruptions can have rippling impacts on the global healthcare system.[7] The pharmaceutical sector is therefore both vulnerable to physical climate impacts and also influential in maintaining public health as the climate changes, making it particularly important for investors to understand how these companies may be exposed to climate risks.

Identifying Risk Exposure

To gain an understanding of exposure to climate hazards in the pharmaceutical industry, we looked at Four Twenty Seven's climate change risk scores[8] for ten of the largest biotechnology companies by market capitalization.[9] This includes companies headquartered in Korea, Japan, the United States, and Ireland, each of which holds global assets. Several of these companies are highly exposed to multiple hazards, with hurricanes, typhoons, and floods presenting substantial risk to many. Some companies also have significant exposure to water stress or heat stress. Such exposure to multiple risks can compound supply chain and operation disruptions when severe events happen simultaneously. Over the past several years, the pharmaceutical sector has experienced the financial consequences of its physical risk exposure, as demonstrated in the following section.

Material Impacts

Hurricanes & Typhoons

Recent history shows that extreme weather events can have material impacts on corporations and sometimes on the market at large if a number of manufacturing sites in a sector are co-located. In the wake of Hurricane Maria in 2017, the United States braced for drug shortages, as manufacturing of cancer, diabetes, and heart disease drugs was reduced and saline bags for medication were already being rationed.[10] Puerto Rico manufactures almost 10% of drugs used by Americans[11] and was severely damaged by Hurricane Maria. While several manufacturing facilities were damaged, companies also struggled to keep supplies cold during enduring power outages and to support employees as homes and infrastructure were damaged. Pfizer Inc. reported a loss of $195 million related to inventory losses and overhead costs when three Puerto Rican manufacturing facilities closed during the hurricane.[12]

As the climate continues to warm, scientists predict an increased number of severe Atlantic hurricanes, even if the impacts on the overall number of storms are less certain.[13] Climate change contributes to slower, rainier storms, which often correspond to more flood damage.[14] Understanding the exposure of critical manufacturing facilities to these events is essential for understanding their business impacts and implementing safety nets for companies and for public health.

Water Stress

India is another critical hub for pharmaceutical manufacturing. Tamil Nadu in Southern India supplies 50% of global demand for several vaccines, with the United Kingdom and the United States as the largest consumers.[15] In the summer of 2019, this region experienced an enduring drought. With three out of four primary reservoirs empty, residents waited in line for water and the costs of purchasing water for surgery continued to rise at Tamil Nadu hospitals.[16] Water stress is a critical risk to pharmaceutical companies because of their reliance on purified water as an input to their products.

Outside of India, water shortages in areas that rely heavily on hydroelectric power, such as Brazil, can also increase the costs of electricity, which can be significant for pharmaceutical manufacturers. Allergan Plc, which has water-stressed sites across the United States, Europe, and India, reported that it has had to deliver water to some water-stressed facilities in the past.[17] This can also increase reputation risk, as communities struggling to access sufficient freshwater witness private trucks delivering water to businesses.

Climate change will contribute to more severe droughts, with negative implications for the global pharmaceutical and biomedical sectors in particular.[18] Operations and supply chain disruptions pose business risks for pharmaceutical companies while evoking concerns about drug shortages and water management in communities around the world.

Heat Stress

Pharmaceutical and biotechnology companies also show significant exposure to heat stress, particularly in the Midwestern United States, India, and Brazil. Pharmaceutical drugs and their ingredients must be stored at precise temperatures to maintain their function. Pharmaceutical companies are therefore sensitive to higher temperatures, which can lead to increased energy costs for cooling, and loss of products if heat waves lead to power outages for which companies are not prepared. Rising global temperatures will exacerbate droughts while also contributing to the increasing severity and duration of heat waves, driving operational costs up, and increasing the risk of stock losses across the industry.

The Need for Improved Physical Climate Risk Disclosures

Reporting on climate-related risks has historically been weak.[19] Financial filings typically include generic language on extreme weather events and catastrophes without providing a detailed, informed view of whether key sites or components of a corporation's value chain are exposed to climate impacts. As climate hazards increasingly lead to financial loss and cascading economic impacts, there is growing concern over the potential financial crisis that may emerge if financial decisions continue to disregard climate risks.[20] Understanding whether corporate risk disclosures accurately reflect firms' physical risks is an important first step in making sure that financial markets adequately understand and price climate change risk.

Using Four Twenty Seven's climate risk dataset as an indication of which pharmaceutical companies have high climate change risk exposure in their operational footprint, we identified if companies disclosed and managed these risks in their financial filings. We reviewed corporate 10-Ks when available and other annual investor reports. We supplemented this research by reviewing companies' CDP reports,[21] which offer more details on corporations' approaches to climate change even when they do not consider it material.

An important characteristic of climate change is that the future will not resemble the past. Thus, firms cannot rely on their past risk exposure to determine if physical climate change risks will be material moving forward. Informative climate risk disclosure combines an understanding of past events' financial impacts with a forward-looking view of climate risk exposure. This information is essential for investors, but also for corporations to manage their risks themselves. However, firms neither disclose the impact of past events consistently nor do they explicitly integrate forward-looking climate projections into their disclosures. Below, we discuss three elements of risk disclosure that corporations can improve to better enable financial markets' pricing of physical climate risks.

Assessment Methodologies

Financial filings do not provide specific information on physical climate risk assessment. Firms tend to frame climate change as a stewardship concern in their financial filings rather than as a material risk. CDP reports contain many more details on exposure and management of climate and water risk, but demonstrate inconsistent terminologies, processes, and methodologies for assessing and reporting such risks. CDP climate and water reports identify and discuss "substantive" risks and opportunities. While materiality is a financial and legal concept consistently used in financial disclosures and explicitly integrated in the Task Force on Climate-related Financial Disclosures recommendations, there is no shared definition of "substantive" in CDP reporting. This reality makes it difficult for investors to identify and minimize material financial risk from climate change.

Most companies do not disclose the process they rely on to identify their climate change risks. This makes it difficult for investors to understand assumptions and time frames and compare relative findings. For example, many companies mention regional trends in climate risk, saying broadly that their facilities are likely to be exposed to and impacted by more climate hazards, or vaguely asserting that despite events becoming more frequent, they aren't expected to impact their business.

Companies must back up such claims with quantitative, forward-looking data if they are to be useful to investors. Allergan, for example, identifies "increased severity of extreme weather events such as cyclones and floods," as the primary climate-related risk driver, noting that the type of financial impact driver is "reduced revenue from decreased production capacity (e.g., transport difficulties, supply chain interruptions)." However, Allergan says that these are potential long-term impacts which are "unlikely."[22] Four Twenty Seven's data, however, shows that Allergan has 15 sites exposed to hurricanes and typhoons, including sites in the United States, the Philippines, Japan, and Taiwan. It would be helpful for investors to understand how Allergan reached its conclusion, so they can compare its self-reported risk against other companies' risk and understand which assumptions and parameters of the risk assessment may be different.

Several firms mention regions where they have facilities exposed to extreme events such as hurricanes, typhoons, and floods, but the granularity of details they provide varies widely. Takeda Pharmaceutical Company specifies that one month without its intermediate Leuplin manufacturing plants would lead to a 9.5 billion Japanese yen loss from reduced sales, or about

$88 million.[23] While disclosing the specific activity and location of sites can be seen as a trade secret, by providing this information, Takeda offers a more transparent picture of its risk exposure and a foundation upon which to discuss resilience measures, such as facility strengthening initiatives taking place at the Leuplin site. However, Takeda mentions this facility as an example of its exposure rather than systematically disclosing all of its exposed sites. Four Twenty Seven's data shows that Takeda has 79 sites exposed to hurricanes and typhoons across Japan, which suggests that it may have additional facilities that are at risk of going offline. Investors should receive information about the potential loss from disruptions at these facilities, as well, because loss from one facility does not necessarily equate to the amount that would be lost by a different type of facility in a different location.

Investors and the market at large both need consistent and comparable information to effectively integrate climate change risk into financial decision-making. To present decision-useful information on climate change risk, companies should disclose their forward-looking risk exposure, explicitly indicating the time horizon they used, which scenarios they explored, and the level of granularity they included. The specific location of corporate assets combined with the most granular climate data that companies can use will offer the most informative disclosures.

Risk Metrics

Translating climate risk exposure into quantitative financial materiality presents significant challenges, as standardized metrics have not yet been developed. Yet, consistent risk metrics coupled with clearly defined terms and replicable indicators will be necessary for investors to compare corporations' physical risk exposure across a portfolio of companies. While research and thought leadership efforts continue to explore these connections and work toward metrics,[24] market leaders are agreeing on quantitative, informative indicators that can communicate exposure. When provided consistently, this information can help investors begin to understand companies' exposure to physical climate hazards and allow climate risks to be more systematically considered by financial markets.

For example, companies can report the percent of their facilities that are exposed to specific hazards, including hurricanes and typhoons, sea level rise, inland floods, heat stress, water stress, and wildfires. This indicates how they may be impacted as specific hazards become more frequent. However, as discussed above, it is essential that companies explain the methodology used to identify the percent of their facilities exposed. Several companies include the number of facilities at risk to current water stress in their CDP reports, but Four Twenty Seven data tends to show many more facilities also exposed to future water stress. For example, Four Twenty Seven's data shows that 17 of Daiichi's sites are exposed to water stress, in India, China, and the United States through 2040. According to its CDP report, Daiichi found that three of its plants are at the "highest risk." Daiichi does mention that these plants together make up less than 5% of company revenue, which offers a helpful metric that puts the potential exposure in context. This disclosure suggests that the risk to these three facilities might not currently be financially material, though it also highlights the potential value of the other exposed sites that may not be the "highest risk" but may still be affected by increasing water stress over time.

Another informative metric would be disclosure of the financial impacts of past extreme weather events to provide some indication of how increasingly frequent and severe events may affect a company's future performance. Bristol-Meyers Squibb, for example, cites the rippling regional impacts of hurricanes on its two manufacturing sites in Puerto Rico, emphasizing that decreased production there may have been caused by several impacts, including facility disruption, infrastructure disruption, reduced services on the island, supply chain disruptions, and staff disruptions.[25] The more specific firms can be in describing how and why they incurred damage, the better investors will understand how their risk is likely to increase over time. For example, as mentioned above, Pfizer reported that it lost $195 million when three Puerto Rico manufacturing sites couldn't operate during the hurricanes.[26]

Lastly, even if companies have not yet completed a comprehensive, forward-looking climate change risk assessment, they can disclose the location of their facilities to enable investors to determine the forward-looking risk for themselves based on regional climate trends or external location-based risk assessments. When disclosing a facility's location, including its activity can provide valuable insight by signaling the sensitivity of these operations to each climate hazard, which lays the groundwork for understanding appropriate asset-level climate risk management. Many pharmaceutical companies consider the precise drugs being produced at specific manufacturing sites to be trade secrets and companies may fear the impacts of first-mover disadvantage, but it is important for companies to gather this information to inform their own climate risk management strategies and to disclose the most reasonable level of detail. As investors increasingly demand more thorough risk disclosure, there is also the potential for companies to be rewarded for increasing their transparency.

Management at Scale

Investors need to understand not only a company's risk exposure but also its management strategies for addressing these risks. Understanding how firms are preparing for climate risks is essential for understanding how risk exposure may translate into material financial impacts and identifying investment opportunities, such as supporting resilience efforts. Companies tend to qualitatively disclose their emergency response efforts for one or two extreme weather events. For example, Eisai Co. Limited's CDP climate report highlights that when Cyclone Hudhud hit Vizag, India, the firm convened its emergency Crisis Management Committee and set up a Crisis Response Headquarters in Tokyo. Though the India facility lost power and water for a week, it incurred no major loss.[27] While a successful emergency response plan is a valuable component of disaster preparedness, this is different from a consistent strategy to identify and manage climate risks at scale and over time.

Extreme events are likely to occur with more intensity and severity in the future and are likely to interact with one another. Hurricanes Harvey, Irma, and Maria, for example, came one after another in 2017. Likewise, extreme heat events often exacerbate enduring droughts. Therefore, ad hoc emergency response efforts based merely on the impacts of past disasters will not suffice for building climate resilience. Companies must instead develop robust, forward-looking plans and integrate preparedness efforts into standard operations, and must disclose these efforts so that investors have a full picture of companies' risk exposure alongside their resilience. Metrics on insurance coverage, insurance cost changes, and uninsured losses, as well as investment into retrofitting facilities or improving resilience at large, are helpful for investors. These metrics are not currently part of financial reports at the level of granularity that would allow an investor to form an opinion on how the company is managing its exposure to climate change. Yet, as disasters continue to occur at a greater frequency, demand for more transparency is expected to increase.

Conclusion: A Need for Climate Risk Standards

While the Task Force on Climate-related Financial Disclosures' recommendations have helped encourage companies to integrate climate change risks into their financial disclosures, the development and implementation of methodologies for meaningfully assessing and disclosing physical climate risk must continue to progress. Investors cannot thoroughly understand their own climate-related risks without receiving informative disclosure from the companies in which they invest. This understanding of physical climate risk exposure is a critical first step in managing risk and supporting resilient economies.

The market has yet to reach consensus on the most informative timeframes and scenarios for assessing climate risk exposure. As corporate and financial actors continue to reflect upon the best way to integrate forward-looking, asset-level climate risk data into their financial materiality assessments, firms must do their best to thoroughly disclose their risks. Companies have an opportunity to get ahead of the competition by describing in detail the time horizons and methodologies they use, which would also promote knowledge exchange and the development of best practices across their respective industries. This in turn will help the financial sector to improve its understanding of the implication of physical climate risks for financial markets and more accurately price this risk.

Notes

1. Law, T. (2019, August 20). Record-Breaking Temperatures Around the World Are 'Almost Entirely' Due to Climate Change. *Time*. Retrieved from https://time.com/5652972/july-2019-hottest-month/.
2. International Labour Organization. (2019). *Increase in Heat Stress Predicted to Bring Productivity Loss Equivalent to 80 Million Jobs*. Retrieved from https://www.ilo.org/global/about-the-ilo/newsroom/news/WCMS_711917/lang--en/index.htm.
3. Koning Beals, R. (2019, November 2). Banks Increasingly Unload Flooded-Out Mortgages at Taxpayer Expense. *MarketWatch*. Retrieved from https://www.marketwatch.com/story/climate-change-could-impact-your-mortgage-even-if-you-live-nowhere-near-a-coast-2019-09-30.
4. Newberry, L. (2018, August 31). As California Fire Disasters Worsen, Insurers Are Pulling Out and Stranding Homeowners. *Los Angeles Times*. Retrieved from https://www.latimes.com/local/lanow/la-me-ln-wildfire-homeowners-insurance-20180830-story.html.

5. Mazzacurati, E. (2017, July 14). TCFD Releases Final Recommendations. *Four Twenty Seven*. Retrieved from http://427mt.com/2017/07/14/final-tcfd-recommendations/.

6. Climate change and environmental data characterizing various geophysical characteristics typically have different units, which can be difficult to compare. It can also be challenging to obtain globally consistent datasets. Climate change data is typically projected in long time frames, beyond the typical business planning horizon. Likewise, the impacts of physical climate change hazards will be based on the exposure of specific facilities, but it can be challenging to obtain granular, location-specific climate change data that is also globally comparable.

7. The Department of Homeland Security. (2018). The Public-Private Analytic Exchange Program. *Threats to Pharmaceutical Supply Chains*. https://www.dhs.gov/sites/default/files/publications/508%20-%20AEP%20Pharmaceutical%20Final%20w-DS%200792018.pdf.

8. Four Twenty Seven leverages global climate change data integrated into a proprietary model to provide asset-level risk assessments of corporations by screening over a million global corporate facilities to score the exposure of publicly listed companies to climate change hazards. The Four Twenty Seven physical climate risk score includes three components: Operations Risk, Supply Chain Risk and Market Risk. Each risk dimension is scored on a scale of zero (least exposed) to 100 (most exposed), and scores are normalized so they are comparable across diverse portfolios. Operations risk is assessed at the facility level. A facility's score for each climate hazard is a combination of the local exposure to the hazard and the facility's specific sensitivity to that hazard. A company's score is an aggregation of each of its facilities' scores. Each climate hazard has underlying indicators to capture the projected change in frequency and severity looking out to mid-century and comparing with a historical baseline of 1975–2005. For each facility, Four Twenty Seven assesses exposure to heat stress, water stress, extreme precipitation, inland and coastal floods, hurricanes and typhoons, using global climate models, NASA data and other environmental data sets.

9. We needed a parameter to narrow down our dataset since we couldn't look at the whole market and chose to look at the largest companies by market capitalization because larger companies tend to have more resources for climate risk disclosures and were thus more likely to have detailed reports for us to review.

10. Thomas, K., & Kaplan, S. (2017, October 4). Hurricane Damage in Puerto Rico Leads to Fears of Drug Shortages Nationwide. *The New York Times*. Retrieved from https://www.nytimes.com/2017/10/04/health/puerto-rico-hurricane-maria-pharmaceutical-manufacturers.html.

11. The Public-Private Analytic Exchange Program.

12. Pfizer Inc. and Subsidiary Companies. (2017). *Appendix A 2017 Financial Report*. http://www.annualreports.com/HostedData/AnnualReportArchive/p/NYSE_PFE_2017.pdf.

13. Rahmstorf, S., Emanuel, K., Mann, M., & Kossin, J. (2018, May 30). Does Global Warming Make Tropical Cyclones Stronger? *Real Climate*. Retrieved from http://www.realclimate.org/index.php/archives/2018/05/does-global-warming-make-tropical-cyclones-stronger/.

14. Kossin, J. (2018). A Global Slowdown of Tropical-Cyclone Translation Speed. *Nature, 558*, 104–107. https://doi.org/10.1038/s41586-018-0158-3.

15. Blodgett, B. (2019, July 26). How India's Drought in Tamil Nadu Affects US Pharmaceuticals. *EDM Digest*. Retrieved from https://edmdigest.com/original/drought-us-pharmaceuticals/.

16. Chaudhary, A., & Kumaresan, S. (2019, July 10). India's Worsening Drought Is Forcing Doctors to Buy Water for Surgery. *The Economic Times*. Retrieved from https://economictimes.indiatimes.com/news/politics-and-nation/indias-worsening-drought-is-forcing-doctors-to-buy-water-for-surgery/articleshow/70155777.cms?from=mdr.

17. CDP. (2018). *Allergen plc. Water 2018*. https://www.cdp.net/en/formatted_responses/responses?campaign_id=62452160&discloser_id=708797&locale=en&organization_name=Allergan+plc&organization_number=533&program=Water&project_year=2018&redirect=https%3A%2F%2Fcdp.credit360.com%2Fsurveys%2Fm88pm39w%2F12807&survey_id=58156493.

18. Schwartz, J. (2018, December 12). More Floods and More Droughts: Climate Change Delivers Both. *The New York Times*. Retrieved from https://www.nytimes.com/2018/12/12/climate/climate-change-floods-droughts.html.

19. TCFD. (2016). *Phase I report*. https://www.fsb-tcfd.org/publications/phase-i/.

20. Fink, L. (2020, January). A Fundamental Reshaping of Finance. *Blackrock*. Retrieved from https://www.blackrock.com/corporate/investor-relations/larry-fink-ceo-letter; Green, M. (2019, September 23). BOE's Carney Urges Financial Sector to Transform Management of Climate Risk. *Reuters*. Retrieved from https://www.reuters.com/article/us-climate-change-un-carney/boes-carney-urges-financial-sector-to-transform-management-of-climate-risk-idUSKBN1W825B.

21. CDP. https://www.cdp.net/en.—A global non-profit offering a disclosure system for investors, companies, and jurisdictions, with questionnaires for climate change, forests, and water security.

22. CDP. (2018). *Allergan plc—Climate Change 2018*. https://www.cdp.net/en/formatted_responses/responses?campaign_id=62255737&discloser_id=763786&locale=en&organization_name=Allergan+plc&organization_number=533&program=Investor&project_year=2018&redirect=https%3A%2F%2Fcdp.credit360.com%2Fsurveys%2Fft9rgfbw%2F12838&survey_id=58150509.

23. CDP. (2018). *Takeda Pharmaceutical Company Limited—Climate Change 2018*. https://www.cdp.net/en/formatted_responses/responses?campaign_id=62255737&discloser_id=653060&locale=en&organization_name=Takeda+Pharmaceutical+Company+Limited&organization_number=18316&program=Investor&project_year=2018&redirect=https%3A%2F%2Fcdp.credit360.com%2Fsurveys%2Fft9rgfbw%2F29440&survey_id=58150509.

24. See in particular: European Bank for Reconstruction & Development (EBRD) and Global Centre on Adaptation (GCA). (2018). *Advancing TCFD Guidance on Physical Climate Risk and Opportunities.* https://s3.eu-west-2.amazon aws.com/ebrd-gceca/EBRD-GCECA_draft_final_report_full_2.pdf.

25. CDP. (2018). *Bristol-Myers Squibb—Climate Change 2018.* https://www. cdp.net/en/formatted_responses/responses?campaign_id=62255737&discloser_ id=736250&locale=en&organization_name=Bristol-Myers+Squibb&organizat ion_number=2191&program=Investor&project_year=2018&redirect=https% 3A%2F%2Fcdp.credit360.com%2Fsurveys%2Fft9rgfbw%2F11745&survey_ id=58150509.

26. CDP. (2018). *Pfizer Inc.—Climate Change 2018.* https://www.cdp.net/en/for matted_responses/responses?campaign_id=62255737&discloser_id=771129& locale=en&organization_name=Pfizer+Inc.&organization_number=14683& program=Investor&project_year=2018&redirect=https%3A%2F%2Fcdp.credit 360.com%2Fsurveys%2Fft9rgfbw%2F25892&survey_id=58150509.

27. CDP. (2018). *Esai Co., Ltd., Climate Change 2018.* https://www.cdp.net/ en/formatted_responses/responses?campaign_id=62255737&discloser_id=702 822&locale=en&organization_name=Eisai+Co.%2C+Ltd.&organization_num ber=5351&program=Investor&project_year=2018&redirect=https%3A%2F% 2Fcdp.credit360.com%2Fsurveys%2Fft9rgfbw%2F17638&survey_id=581505 09.

6

Mosaic Theory in Sustainable Investing

Valerie S. Grant

Abstract *Mosaic theory*—the process of synthesizing information from many sources to build conviction about a company's prospects, valuation, or credit-worthiness—has been a central tenet of investing for decades. The emergence of sustainable investing, combined with the ability to access large volumes of structured and unstructured data, has amplified the importance of mosaic theory as well as its complexity. Investors must overcome the natural tendency of a company's management team to highlight favorable information in corporate disclosures and for auditors and regulators to focus on controls and compliance. Corporate disclosures related to environmental, social, and governance issues should be compared to other relevant data sources to assess a company's prospects, particularly in the context of evolving standards of corporate conduct from shareholders, employees, customers, and other stakeholders.

Keywords Mosaic theory · Sustainable investing · Non-financial disclosure · Materiality · ESG data · Data diversity · ESG metrics · Responsible investing · Non-public data · Long-term risks · Strategic vulnerability · ESG disclosure

V. S. Grant (✉)
AllianceBernstein LP, New York, NY, USA
e-mail: valerie.grant@alliancebernstein.com

© The Author(s) 2020
D. C. Esty and T. Cort (eds.), *Values at Work*,
https://doi.org/10.1007/978-3-030-55613-6_6

This chapter explores the benefits—and limitations—of company disclosures related to financial and non-financial performance indicators. It advocates for the continued relevance of *mosaic theory* as a foundation for sustainable or responsible investing. Finally, it emphasizes the role that investors—armed with their mosaics—can play in motivating companies to improve their performance.

Mosaic theory is an approach to investment research that relies on gathering, analyzing, and drawing conclusions from public, non-public, and non-material information. Widely used by research analysts and portfolio managers, it helps investors to determine the value of a company's stock or the likelihood that it will default on its credit obligations. This approach is more relevant than ever in the burgeoning field of sustainable investing, where data is fragmented, disclosures are uneven, and the relevance of different pieces of information is in the eye of the beholder. In today's market, investors can take advantage of management disclosures and alternative data sources like online reviews, web-traffic information, and the visualization of physical activities to build a nuanced and more complete view of a company's prospects.

As sustainable investing becomes widely adopted by individuals and institutions who want their investments to align with their values without sacrificing risk-adjusted returns, mosaic theory offers an important tool for helping to understand how companies are performing and whether their performance on environmental, social, and governance (ESG) issues is mispriced. Certainly, improved disclosures on ESG issues will give investors more transparency and timely information on key performance indicators. Investors must, however, be cautioned against putting too much emphasis on any set of established standards for disclosure, because even the most clearly defined standards can be interpreted differently—and even manipulated—by preparers of corporate disclosures. The obstacle to better disclosure of material ESG issues is not a lack of understanding or a lack of clarity on what is or is not material. Rather, it is the natural reluctance of management teams to be forthcoming about opportunities and risks that are difficult to predict, longer term in nature, or which may highlight areas of company exposure or strategic vulnerability.

If investors have learned anything from the long history of using financial disclosures, it is that well-articulated standards for disclosure are necessary, but not sufficient, to make investment decisions and allocate capital, for three important reasons:

- Incentives that influence management's willingness to disclose comprehensive and balanced financial information are different from those of auditors and users of disclosures.

- Preparers, auditors, and users of ESG-related disclosures also face different, and often conflicting, incentives.
- Analyzing information that companies disclose is only one of many steps needed to make sound, responsible investment decisions.

The analysis begins by exploring the incentives that motivate the preparers, auditors, and users of company disclosures.

Incentives for Comprehensive Disclosure May Not Align

Those within companies who prepare corporate disclosures, outside auditors, regulators, and investors all have an interest in what companies disclose. But because each of these interested parties has access to different information with different rewards—or punishments—for different outcomes, their interests fail to align.

Senior executives interpret accounting principles and prepare corporate disclosures in a way that prioritizes the maintenance of a company's access to capital and, to varying degrees, the maximization of their compensation. Lawyers at public companies, though well-meaning, review disclosures before they are issued, often recommending broad, boilerplate language that addresses the issues in question but is not specific enough to be fully useful to other parties. Executive teams, even those whose compensation plans attempt to align their interests with those of investors, often lack the incentive to disclose activities, technologies, or risk assessments that entail uncertainty or that have a long time horizon.[1] For these reasons, they generally focus disclosure on favorable information and information with a short time horizon.

Auditors, whether accountants or representatives of regulatory agencies, inspect and verify financial and ESG filings. They seek to confirm that the statements prepared by management conform to Generally Accepted Accounting Principles (GAAP) and abide by the law. Their primary incentive is to detect fraud and violations of securities laws or regulatory guidelines, and they are motivated by a desire to protect their own enterprises from both legal liability and reputational damage resulting from auditing the books of a company that may later be found to have acted improperly. A "clean" or unqualified auditor's report concludes that financial statements fairly present the results of a company's operations based on generally accepted accounting standards. It also confirms the presence of adequate checks and balances, known as internal controls. The report does not necessarily reveal all material risks, nor does it point out commercial opportunities that might improve future cash flows or benefit society. Those responsibilities require interpreting financial statements and projecting future cash flows, tasks that lie squarely within investors' scope of responsibility.

Investors and creditors are the main users of financial statements—and increasingly of ESG reports. Those in the investment arena evaluate the information disclosed by management and analyze it in the context of other information and data they obtain about the drivers of a company's future cash flows. Investors then determine whether the company is under or overvalued or—from the perspective of creditors—likely to repay its debts. As individuals and institutions increasingly seek to balance return, risk, and corporate responsibility, investors and creditors are also allocating capital through the lens of a company's positive or negative impact on society.

Other stakeholders—including employees, pensioners, government regulatory agencies, customers, and suppliers—also use financial and non-financial disclosures. As responsible investing becomes more widely adopted among individuals and institutions, the interests of these stakeholders are converging with those of investors and creditors. These parties are all motivated by the desire for, and often frustrated by the lack of, comprehensive, balanced information about an enterprise's financial and operating performance and its impact on society.

Unlike in private markets, where investors have unconstrained access to material, non-public information, investors in public markets must rely on management disclosures: financial statements, explanatory notes, earnings calls, and other notices regulators require. In practice, these disclosures include only some of the information available to management. Management teams have broad discretion in interpreting standards of materiality and applying accounting principles.

To overcome this information gap, investors incur costs to investigate and monitor what management teams are doing: How are they allocating capital? Are employees complying with the letter and spirit of laws and regulations? Has management entered into related-party agreements that put their interests ahead of those of shareholders and other stakeholders?[2]

Given this conundrum, investors must seek information from sources other than disclosures to validate the data provided by management. Even when disclosures are robust, it is prudent to apply a "trust, but verify" approach to securities analysis.

Case Study 1: Impact of Accounting Standards Update 2017-07 on the U.S. Consumer Packaged Food Sector

In 2017, the Financial Accounting Standards Board (FASB) issued Accounting Standards Update No. 2017-07, *Compensation-Retirement Benefits (Topic 715): Improving the Presentation of Net Periodic Pension Cost and Net Periodic Postretirement Benefit Cost*.[3] The update was designed to give investors more clarity about costs associated with companies' pension plans and other retirement benefits.

Before the change, management teams in the U.S. consumer packaged food industry commonly measured corporate profitability in conformance with Generally Accepted Accounting Principles (GAAP) standards, supplemented by measures based on internally defined standards ("non-GAAP"). Disclosing GAAP and non-GAAP measures of profitability is permissible under securities laws if non-GAAP measures are clearly labeled as such. For GAAP Earnings Before Interest and Taxes (EBIT), a commonly used measure of corporate profitability, firms could include five distinct metrics related to pension income and/or expense in the calculation: service cost,[4] interest cost, expected return, actuarial gains and losses, and other one-time items. The new guidance was less flexible—only service cost could be included. The other adjustments had to be excluded from GAAP EBIT, although they could be included in other parts of the income statement. This more narrowly defined measure of profitability substantially reduced the profit margins of several companies, including Kellogg, Kraft Heinz and ConAgra.[5]

Management teams in this industry also reported supplemental profitability measures that did not conform to GAAP. "Adjusted EBIT" was often disclosed alongside "GAAP EBIT," even though assumptions underlying adjusted EBIT varied widely. Because this was a non-GAAP measure, management teams had much more latitude to determine what to include or exclude. The variances in adjusted EBIT following the new accounting guidance were somewhat more muted. However, once again, the profit margins of Kraft, Kellogg, and ConAgra were materially worse once they adjusted their reporting to align with the new accounting guidelines.

Companies in this sector were already on shaky ground when the guidance was issued. Sales volumes and profit margins were under pressure, many brands were losing market share to private-label products and innovative health-conscious brands, and organizations were struggling to respond to shifts in consumer preferences favoring fresh and premium prepared foods over processed mass-produced products.[6] These companies could not pass through price increases, and creatively accounting for pension liabilities on the income statement was one way to mask the problem.

Before Financial Accounting Standards Board (FASB) Standard 2017-07 went into effect, price-to-forward earnings multiples[7] traded in lockstep with each other. However, once investors analyzed firms under the new accounting guidelines, they began to differentiate among companies in the industry. Valuation multiples increased for McCormick, Hershey, and Mondelez—firms whose profitability was higher and more stable than previously believed—and decreased for Kraft Heinz, ConAgra, JM Smucker, and Kellogg. The laggards (except for JM Smucker) all disclosed lower EBIT and adjusted EBIT margins after FASB revised its reporting guidelines.

Given the confluence of events when the accounting guidance and related financial disclosures changed, it is impossible to know whether that specific piece of information caused multiples to decline for the "bad actors." However, the added disclosures were certainly another piece of information in the *mosaic* that supported a more pessimistic outlook for these stocks.

Incentive Misalignments in ESG Disclosures

Some academicians and practitioners, having learned from the deficiencies of financial disclosures, advocate for a more prescriptive approach to sustainability reporting standards. The focus on sector- or industry-specific materiality guidelines is driven by a desire to provide guidance to management on how to measure and report on sustainability performance. Advocates also want to give investors and other stakeholders access to comparable performance measures between companies and over time.[8] Advocates implicitly assume that if management teams *knew* better, they would *do* better. In other words, a lack of understanding has led to a lack of disclosure.

This reasoning is flawed. Self-interest is the obstacle to better ESG disclosures, as is the case with traditional disclosures. The incentives that guide the behavior of preparers, auditors, and users of ESG disclosures are not aligned. The challenge is further complicated by the fact that many ESG disclosures are not audited. In responsible investing, the incentives that motivate behavior vary widely, and relevant information is not equally available to all stakeholders with an interest in ESG disclosures. This group includes investors, regulatory agencies, advocacy groups, suppliers, customers, employees, and policymakers.

Having prepared and used financial and non-financial disclosures, I have a sense for the wide gap between the information available to management teams and the information disclosed in securities filings. Most senior leaders and financial officers have good intentions and seek to provide balanced information, but investors cannot underestimate the subtle pressure on management teams to present information in the most favorable light. After all, that information often directly influences a company's stock price, access to credit, and thresholds for incentive compensation.

With sustainability reporting, companies retain an incentive to focus disclosure on favorable information. They seek favorable ESG ratings just as they seek favorable credit ratings in bond markets and "buy" recommendations from sell-side analysts in equity markets. In a study from the Sustainability Accounting Standard Board (SASB), companies failed to report 90% of the known negative ESG events that involved them. This happens in part because sustainability reporting is voluntary.[9] Some academics and policymakers have advocated for mandatory ESG disclosures.[10] While this would certainly help, it is unclear that disclosures alone will offer practitioners the insight that they need to forecast a company's future cash flows, determine whether the full range of outcomes has been "priced in," and build portfolios that deliver on client's expectations for return, risk, and ESG characteristics.

Case Study 2: Social Indicators and the Demise of For-Profit Colleges

In the vortex of the Global Financial Crisis in 2008, the for-profit education sector was one of few areas of the stock market to rally. Apollo Education Group, Strayer, DeVry, ITT Educational Services, Corinthian Colleges, and Career Education Corporation were among the leaders at the time. As the unemployment rate rose, displaced workers went back to school to upgrade their skills and earn degrees and certificates in high-demand fields. Enrollment in these programs surged as community colleges and traditional universities failed to meet the growing demand.

The Allure of For-Profit Colleges

The growth of for-profit colleges sparked controversy then, as it does now, due to the industry's reliance on federal financial aid, the quality of their programs, and the high default rates of their graduates and former students. Advocates of this sector argued that for-profit colleges offered non-traditional

college students an opportunity to earn college degrees or relevant professional certificates, while critics argued that the schools used high-pressure
sales tactics to enroll students in costly programs that were de-linked from
the labor market, ultimately saddling students with loans they could never
repay.

For-profit colleges were originally designed to offer educational services
and credentials to working adults with some college experience or equivalent
work experience. Over time, they expanded to offer associates and bachelor's
degrees to recent high school graduates and adults with limited preparation
for college-level coursework. Students appreciated for-profit colleges' capacity
and flexibility in enrollment and curriculum delivery. Education advocates
saw an avenue for closing persistent "skills gaps" in local labor markets.
Investors were attracted by the colleges' profitable business models, which
they perceived as durable and scalable with compelling growth characteristics.

In addition to standard financial statements, for-profit education companies disclosed information on classroom-based and online "new starts" and
"total enrollment" by program. Those firms offering online programs usually
disclosed those enrollment figures, too. However, they did not release information on graduation rates, starting salaries, or job placement rates. They
did disclose bad debt expense, as measured by U.S. Generally Accepted
Accounting Principles (GAAP) on the income statement. Bad debt expense
indicates the capacity and willingness of students to pay tuition expenses. The
U.S. Department of Education also released information on cohort default
rates for each school, which it used to determine whether enrolled students
were eligible for federally guaranteed student loans and Pell grants.

Problems Brewing

As enrollment surged, industry critics and regulators at the Department
of Education grew more vocal, citing high-pressure recruiting tactics and
complaints of lower-than-expected education quality. As the economy slowly
improved, some investors began to challenge the value proposition of these
companies and to request broader disclosure for three reasons: (1) a spike in
default rates for pools of student loans that had been sold to investors as safe
investments with strong credit characteristics; (2) rumors that new Department of Education regulations would require greater disclosure of graduation
rates, salaries, and job placement statistics; and (3) a proposal from Department of Education to use three-year rather than two-year default rates on
student loans in determining programs' eligibility for federal funding.

Initial drafts of the regulations required schools to demonstrate that they
were preparing their graduates for "gainful employment," which would be

based on post-graduation income relative to student loan burden. In 2010, the U.S. Government Accountability Office issued written testimony entitled "For Profit Colleges: Undercover Testing Finds Colleges Encouraged Fraud and Engaged in Deceptive and Questionable Marketing Practices."[11] Prime-time program *60 Minutes* covered the report soon after.

Bulls vs. Bears: Who Was Right?

Company management teams resisted the regulations, defended the industry's value proposition to students, and claimed that for-profit schools were being singled out unfairly. Critics cited graduates' high debt burdens and dependence on federally guaranteed student loans as a justification for tighter regulation.

Meanwhile, some investors bought the stocks, confident that the regulatory risks would abate, while others concluded that greater transparency and government oversight would ultimately erode the profit margins of for-profit colleges and erode their market value. The buyers and sellers of these stocks were all frustrated by the lack of disclosure from management teams and began to broaden their information mosaics. This expanded perspective included analyzing pools of loans that had been issued by Sallie Mae, the largest issuer of federally guaranteed student loans at the time. Investors also analyzed online search activity for indications of changes in demand for educational services, and interviewed students, recent graduates, and former employees of for-profit colleges.

Analyzing Material Information: Necessary But Not Sufficient

Some members of the responsible investing community are fixated on defining materiality, a term with varying interpretations. According to the Financial Accounting Standards Board, "general purpose" financial reporting should:

> provide financial information about the reporting entity that is useful to existing and potential investors, lenders, and other creditors in making decisions about providing resources to the entity. Those decisions involve buying, selling, or holding equity and debt instruments and providing or settling loans and other forms of credit.[12]

The International Accounting Standards Board recently revised and clarified its definition of *material*:

> Information is material if omitting, misstating or obscuring it could reasonably be expected to influence the decisions that the primary users of general-purpose financial statements make on the basis of those financial statements, which provide financial information about a specific reporting entity.[13]

The Sustainability Accounting Standard Board uses the definition of materiality established under U.S. securities laws: information is material if there is "a substantial likelihood that the disclosure of the omitted fact would have been viewed by the reasonable investor as having significantly altered the 'total mix' of information made available."[14]

It is laudable to get companies to disclose material ESG information. However, investors should interpret these disclosures—whether made in corporate social responsibility reports, earnings calls, or management meetings—with the same skepticism and rigor they apply to financial disclosures. The quality of information varies by company and can change based on the composition of the leadership team. While some executives are direct and forthcoming when engaging with investors, others minimize the importance of bad news, hoping to delay negative investor reaction.[15] Others bury negative information in footnotes or other less widely read sections of disclosures. Deloitte—one of the "Big Four" accounting organizations—calls this "obscuring" rather than "omitting" material information, which can be equally as effective in undermining the usefulness of financial reports and company disclosures.[16] Finally, investors must consider ESG issues, whether disclosed by the company or surfaced from other sources, in the context of other drivers of company performance and valuation.

Mosaic theory depends on using material and non-material information to develop investment insight (Fig. 6.1). Investors can obtain some information only from management disclosures but can derive more insight than ever by analyzing third-party data sources including online reviews, web-traffic information, trading flows, visualization of physical activities, and government databases. Analyzing and querying these large datasets are a boon for analysts and portfolio managers seeking to build portfolios with specific ESG characteristics. Using alternative data sources to supplement management disclosures, investors can define which metrics they view as most important, control how they are derived, and then compare metrics across companies.

Recently, the chief executive officers of 181 companies signed a new Statement on the Purpose of a Corporation, moving away from shareholder primacy and voicing a commitment to the interests of a broader range of

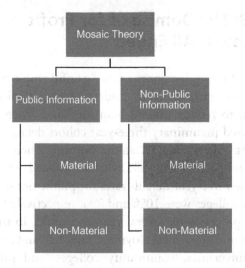

Fig. 6.1 Mosaic theory in traditional investing (*Source* Corporate Finance Institute, "Mosaic Theory: The Analysis of Public and Non-Public Material and Non Material Information to Determine a Security's Underlying Value," www.corporatefinanceinsti tute.com accessed October 18, 2019)

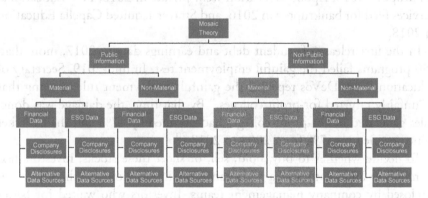

Fig. 6.2 Mosaic theory in context of responsible investing and big data (*Source* Corporate Finance Institute, "Mosaic Theory: The Analysis of Public and Non-Public Material and Non Material Information to Determine a Security's Underlying Value," www.corporatefinanceinstitute.com accessed October 18, 2019; AllianceBernstein analysis)

stakeholders.[17] At the World Economic Forum in Davos, Switzerland in 2019, CEOs from companies as diverse as Credit Suisse, Nasdaq, Cisco, and McKinsey argued for a shift in the conversation about corporate purpose and sustainability.[18] This change in focus means it is no longer enough to build a mosaic that includes only financial issues. As a result, implementing the mosaic theory is now even more complex, as illustrated in Fig. 6.2.

Case Study 3: The Demise of for-Profit Colleges—How It All Ended

As colleges prepared to implement the new "gainful employment rule," many began to cancel programs, invest in student retention initiatives, and offer more financial aid to reduce student loan burdens. When the Department of Education released preliminary three-year cohort default rates for FY2007 in 2009, for-profit colleges ranked at the bottom. Their three-year default rates were 21% compared with 11% when calculated on a two-year basis. By comparison, the three-year default rates for public non-profit colleges and private non-profit colleges were 10% and 7%, respectively.[19]

The regulatory process took several years to unfold. In the meantime, the economy began to improve, unemployment rates fell, and student enrollment declined. State universities, community colleges, and private universities began offering Massive Open Online Courses ("MOOCs") and other educational services, either for free or at very affordable prices. As profit margins contracted, the for-profit industry consolidated. Corinthian College went bankrupt in 2015, Apollo Education went private in 2016, ITT Educational Services filed for bankruptcy in 2016, and Strayer acquired Capella Education in 2018.

In the first release of student debt and earnings data in 2017, more than 750 programs failed the gainful employment test. In mid-2019, Secretary of Education Betsy DeVos repealed the gainful employment rule, arguing that it "unfairly targeted for-profit colleges." By that time, the damage was done. Sales generated by for-profit colleges had declined by 47% and their market value declined by 67% from 2008 to 2018.[20]

To decide whether to buy, hold, sell, or short these stocks, investors had to mobilize and synthesize a wide range of information, much of it never disclosed by company management teams. Investors who waited for better disclosures on key performance indicators on student outcomes and other measures of educational performance missed their opportunity to allocate capital profitably in this volatile market. Short sellers generally fared well.

What Investors Can Do

While most asset owners and asset managers focus exclusively on their portfolios' financial performance, many also want to guide investments with the added objective of managing externalities and/or making a positive impact on society. A rigorous implementation of mosaic theory yields benefits for these market participants, particularly as we confront issues like climate change, income inequality, technological disruption, and changes in the structure of labor markets.

To build an information mosaic, investors must differentiate between data and noise with greater speed, accuracy, and efficiency. Applying the mosaic theory in the context of responsible investing can help offset the lack of full alignment between the interests of preparers, auditors, and users of company disclosures. By identifying the key investment controversies for a stock and then assembling material and non-material information from a broad range of sources, investors can develop greater conviction about the long-term sustainability of a company's cash flows. In addition, by evaluating companies holistically, we are less likely to be blindsided by issues that we might overlook if we rely solely on company disclosures or material information.

Creating a mosaic often involves applying multiple layers of paint or other material to a canvas or other structure. Think of corporate disclosures as the first layer of the information mosaic. It provides a strong foundation, but the real insight will require investors to engage thoughtfully and proactively with management teams, analyze disclosed information provided by the company, validate that information from other sources, and then encourage management teams and directors to consider the long-term implications of their attention to ESG issues in corporate strategy and business operations.

The most beautiful mosaics are not necessarily the most complex. They are the ones that create an object of beauty from otherwise mundane materials and develop recognizable patterns based on the artist's diligence and creativity. For practitioners of responsible investing, company disclosures and big data are our stones and glass: it is up to us to put the pieces together to determine which companies are worthy of our capital and which are not.

Notes

1. Jensen, M. A., & Meckling, W. H. (1976). Theory of the Firm: Managerial Behavior, Agency Costs and Ownership Structure. *Journal of Financial Economics, 3*(4), 305–360.
2. Jensen et al.
3. Financial Accounting Standards Board. (2017). Improving the Presentation of Net Periodic Pension Cost and Net Periodic Postretirement Benefit Cost. *FASB.* Retrieved from https://fasb.org/jsp/FASB/FASBContent_C/Completed ProjectPage&cid=1176168898401.

4. Service cost refers to the present value of the projected retirement benefits earned by plan participants in the current period. Generally, a company's pension service cost is the amount it must set aside in the current period to match the retirement benefits accrued by plan participants.

5. Based on an analysis of company disclosures for FY2017 or FY2018 based on the reporting period in which the accounting change was implemented, which varies by company. Source: company disclosures, AB Bernstein and AllianceBernstein.

6. Howard, A., Ma, Z., & Han, Y. (2018, March 20). *The Long View: US food— What Goes Up Must Come Down; Could EBIT Margins Fall from 19% to 15% Over the Next Decade?* AB Bernstein.

7. The price-to-forward earnings multiple (forward P/E) is the ratio of a company's stock price to its forecasted earnings per share. It is an indicator of how much investors are willing to pay for a company's future earnings. Source: Investopedia, AllianceBernstein LP.

8. Eccles, R. G., Krzus, M. P., Rogers, J., & Serafeim, G. (2012). The Need for Sector-specific Materiality and Sustainability Reporting Standards. *Journal of Applied Corporate Finance, 24*(2), 65–71.

9. Boiral, O. (2013). Sustainability Reports as Simulacra? A Counter-account of A and A + GRI Reports. *Accounting, Auditing & Accountability Journal, 26*(7), 1036–1071.

10. Esty, D. C., & Karpilow, Q. (2019). Harnessing Investor Interest in Sustainability: The Next Frontier in Environmental Information Regulation. *Yale Journal on Regulation, 36*(2), 625–692.

11. Kutz, G. D. (2010). *For-profit Colleges: Undercover Testing Finds Colleges Fraud and Engaged in Deceptive and Questionable Marketing Practices.* United States Government Accountability Office. Retrieved from https://www.gao.gov/assets/130/125197.pdf.

12. Financial Accounting Standards Board. (2018). *Conceptual Framework for Financial Reporting* (Statement of Financial Accounting Concepts No. 8).

13. International Financial Reporting Standards Foundation. (2018). *IASB Clarifies Its Definition of 'Material'.* IFRS. https://www.ifrs.org/news-and-events/2018/10/iasb-clarifies-its-definition-of-material/.

14. Sustainability Accounting Standards Board. (2017). *SASB's Approach to Materiality for the Purpose of Standards Development* (Staff Bulletin No. SB002-07062017).

15. Dizi, A. (2017, May 25). How to Listen for the Hidden Data in Earnings Calls. *Chicago Booth Review.*

16. Deloitte. (2015). *Materiality: Judgements Around Financial Report Disclosure.*

17. Business Roundtable. (2019, August 19). Business Roundtable Redefines the Purpose of a Corporation to Promote 'An Economy That Serves All Americans.' Retrieved from https://www.businessroundtable.org/business-roundtable-redefines-the-purpose-of-a-corporation-to-promote-an-economy-that-serves-all-americans.

18. Birkinshaw, J. (2020, January 29). Why Sustainability Was the Star at Davos 2020. *Forbes*. Retrieved from https://www.forbes.com/sites/lbsbusinessstrateg yreview/2020/01/29/why-sustainability-was-the-star-at-davos-2020/#716a99 0876a3.

19. Finaid. (2020). www.finaid.org. Accessed on September 17, 2019. US Department of Education, AB Analysis.

20. Based on an analysis of data is for FY2008 and FY2018 for all companies that include Apollo Education, Adtalem Global Education (formerly DeVry), ITT Educational Services, Strategic Education Inc. (formerly Strayer), Career Education Corporation, Corinthian College, Capella Education (now owned by Strategic Education Inc.), American Public Education Inc, Universal Technical Institute, and Lincoln Education. FY2018 data for Apollo Education based on estimates and private market valuation as of May 2016. Source: Bloomberg, AllianceBernstein analysis.

7

Toward a Next Generation of Corporate Sustainability Metrics

Daniel C. Esty and David A. Lubin

Abstract Marketplace evidence suggests that a significant number of main-stream *value investors* want to understand which companies are positioned not just to navigate sustainability issues—including climate change, air and water pollution, racial injustice, workplace diversity, structural inequality, privacy, corporate integrity, and good governance—but which will thrive and gain competitive advantage as a result of the emerging *sustainability imperative*. As they did with prior business megatrends, such as the rise of information technologies and globalization, these investors want insights on the strategies developed by management teams to transition their business models for success under profoundly changed societal and market demands. This perspective requires a next generation of high-resolution ESG metrics based on granular, enterprise-wide asset-, process-, and product-level data that illuminate the critical dimensions of corporate performance required to deliver financial success in the face of the sustainability imperative.

Keywords Sustainable investing · Value investors · Business megatrends · Sustainability imperative · Competitive advantage · Business model transitions · Next-generation ESG metrics · Granular analysis · Maturity curves · Data trends · Strategy mapping · Value driver model

D. C. Esty (✉)
Yale University, New Haven, CT, USA
e-mail: daniel.esty@yale.edu

D. A. Lubin
Constellation Research and Technology, New York, NY, USA

© The Author(s) 2020
D. C. Esty and T. Cort (eds.), *Values at Work*,
https://doi.org/10.1007/978-3-030-55613-6_7

The year 2020 will be seen as a watershed moment for many reasons—the COVID-19 pandemic, an unprecedented economic collapse, the Black Lives Matter protests, rising concerns about structural inequality, urgent demands for racial justice, and growing awareness that climate change might track the pandemic's path of chaos and destruction.[1] It may also be remembered as the moment when the world began to take seriously what we have called the *sustainability imperative*.[2] Political leaders announced their determination to "build back better,"[3] and companies across every sector committed to address the full spectrum of sustainability issues from climate change and other environmental challenges, to workplace diversity and corporate governance.[4] Investors have followed suit and poured money into sustainability-oriented funds, signaling their growing interest in holding shares in companies positioned to lead the way toward a future that is decarbonized, economically more fair, and with opportunities fully open to all.[5]

Equally significantly, investment analysts have started to question the environmental, social, and governance (ESG) metrics available to them—and to cast doubt on whether the information being offered about corporate sustainability performance provides sufficient accuracy, comparability, and insight to merit reliance in their stock picking and portfolio construction. It is widely understood that many of the ESG indicators available today are methodologically flawed and do little to differentiate real sustainability leadership from carefully curated corporate narratives that often amount to greenwash when the veneer is scraped away.[6] But when the history of rising interest in sustainable investing is written, we think 2020 will be the year that investors said *enough is enough*—and the push toward a more reliable and instructive set of ESG metrics gained momentum.

Fundamentally, sustainability-minded investors seek a clear line of sight on which enterprises are poised to lead the way toward a sustainable future—and which are not likely to keep up. As Chapter 4 of this volume ("Creating Investment-Grade Corporate Sustainability Metrics") spells out, investors want more carefully curated, analytically rigorous, methodologically consistent, cross-company comparable, externally validated—and thus trustworthy—ESG metrics. But our research suggests that a growing subset of those committed to sustainable investing want something beyond just cleaner and more reliable sustainability data.[7] Specifically, those who not only seek to put their *values* into their investments, but also expect their ESG screening to deliver financial out-performance, want new and different metrics that provide a signal about the capacity of corporations and their management teams to profitably navigate the megatrends that lie behind the *sustainability imperative*.[8] In the past, these *value* investors wanted insights on which companies were prepared to handle previous business megatrends such as quality, technology, and globalization and emerge at the front of the competitive pack. Today, their interest extends to the various dimensions of sustainability.

Elements of the Sustainability Imperative

In this regard, we see a series of sustainability-related challenges firms must address to emerge as industry leaders and stock market successes in the years ahead. These megatrends, which in many cases require significant recalibration of existing business strategies and assumptions, include[9]:

- Deep decarbonization, which implies a shift in the energy foundation of the economy away from fossil fuels toward clean and renewable energy sources;
- A transformed understanding of a corporation's mission—moving away from *shareholder primacy* to *stakeholder responsibility*,[10] which makes unacceptable the business models that privatize gains while imposing costs on the community or burdens on the environment, and will mean the "end of externalities"[11];
- New public expectations related to *racial justice* and *economic fairness* that promise to profoundly reshape the diversity and inclusion agenda in the workplace and across society more broadly;
- Pressures to address structural economic inequality and pay equity; and
- Changes in legislation, regulation, and societal expectations related to corporate lobbying and issue advocacy, integrity, transparency, protection of privacy, and governance.[12]

As with prior megatrends, companies that successfully internalize the various demands of the sustainability imperative and build these new requirements into their business models will be positioned for enduring competitive advantage. Those who fail to transition their products, services, and business practices face greatly increased risks to revenues and reputation. Adapting a business, especially a large complex enterprise, to these disruptive forces requires a corporate journey that will often take many years to see meaningful results, but the progress on this journey can be mapped and measured. Indeed, the authors previously have spelled out how *maturity curves* can be constructed to track how companies are doing in transforming their business models[13]—with sustainability being the latest megatrend that might benefit from such analysis as a way to highlight which companies will thrive and which will fall by the wayside.

Value Driver Model

As the management literature makes clear, thriving companies deliver value through the successful execution of business strategies that are regularly updated and occasionally fundamentally overhauled to stay ahead of changing customer desires, shifting marketplace realities, and evolving societal expectations. Value creation derives, in particular, from corporate capacity for durable revenue *growth*, increasing operating *productivity*, with reduced business *risk*. Various elements of the sustainability imperative now intersect with—and in some cases drive—corporate strategies, creating both requirements for change and opportunities for innovation. Thus, a business strategy-centric view of sustainability would cause an analyst to want to know how a company's sustainability initiatives have been integrated with its core business strategy to deliver material sustainability-related impacts on the key value drivers of growth, productivity, and risk.[14]

With this *value driver* model[15] in mind, we argue that investors need a new structure of ESG metrics that track the *maturity* of a company's management in delivering and executing its sustainability transition strategy, as well as the company's *momentum* in delivering the results required to succeed under changed circumstances. Unlike the current structure of ESG data, the underpinning for these *next-generation* sustainability metrics comes from the world of corporate performance management rather than issue advocacy. We see an analogy to the business analytics of past decades that helped investors assess where companies stood on quality, technology, and globalization maturity curves alongside metrics that captured the direction and rate of change of their business models to capitalize on these megatrends.[16]

These analytic tools helped business leaders understand not only *best practices* on key issues, but also the *next practices* that they must master for continued progress up the maturity curve that tracks the transition required by the megatrend. Such dynamic and forward-looking analyses provide a way to gauge where a company is going, which is much more valuable than a grade on where it is today. This type of granular assessment provides the often-missing context required to make sense of ESG ratings and establish a more meaningful scorecard for the companies being rated.

High-Resolution Analysis

The longitudinal maturity curve analyses we think would be useful requires company-specific assessments based on *empirical, high-resolution*, and *enterprise-wide asset-, process-, and product-level data* that illuminate the critical dimensions of performance required to manage all of the various elements of the sustainability imperative. While today's corporate citizenship, environmental, or sustainability reports may contain numerous examples of reduced pollution, resource efficiency gains, or charitable activities, very few of them offer enterprise-wide aggregated product-, production-, or process-level data that reflect meaningful change on the most significant sustainability challenges. To take sustainable investing to the next level, this gap must be filled with a new framework of high-resolution, strategy-focused ESG metrics.

In the auto sector, for example, approximately 90% of the greenhouse gas emissions derive from the use of the product—vehicles—not production. Thus, the critical decarbonization factor to track will be so-called *Scope 3* emissions emanating from product use, not the *Scope 1* and *2* emissions from manufacturing. But the *Scope 3* emissions reporting available today does not provide the relevant metrics properly normalized. Differences in assumptions on factors such as vehicle life or use patterns greatly distort the company-by-company results. What is therefore needed are granular product-level metrics, such as average grams of CO_2e per vehicle mile of the actual fleet sold. This datapoint, along with information on the change in the vehicle emissions trajectory over time, represents the sort of new metric that would offer investors real insight—especially those who seek *green alpha* based on sustainability-derived marketplace success. Likewise, detailed measures revealing R&D spending on alternative drive vehicles, specific model release plans and sales projections for advanced vehicles, and other production and planning metrics would provide a more meaningful basis of comparison of the transition readiness of major automakers than anything in the ESG data marketplace today. High-resolution data on critical issues and trends are what is expected in mainstream investment analysis—and we think that the next generation of ESG metrics will require as much.

In a similar vein, Michael Porter, in a highly regarded 1996 article,[17] presented *activity mapping* as his method of representing corporate strategy and competitive advantage. He depicted in this piece the six key drivers of Southwest Air's strategy to gain competitive advantage by being *the low-fare airline*. Porter's map of Southwest's strategy highlighted the airline's focus on short-haul flights with limited passenger services, at low ticket prices with quick turn-around departures by highly productive crews delivering industry-leading asset utilization. For Southwest's business strategy to succeed, the company needed, among other things, to achieve higher levels of productivity than their competitors. Accomplishing this goal required a more sophisticated (what we might call more *mature*) human capital strategy that involved higher levels of gate and ground crew compensation to attract and develop the best people as well as stock ownership to drive employee engagement and retention. Though Southwest's people were paid more, the system cost less on a per passenger mile to operate. Although not discussed as such at the time, this strategy, powered by Southwest's human capital investment, resulted in a major success on the social pillar of sustainability and simultaneously delivered a major payout for stakeholders.

Similarly, the profitability of Southwest's low-fare strategy depended upon reducing operating costs. Efficiency came from flying its airplanes efficiently and standardizing its aircraft fleet to reduce maintenance complexity. Not surprisingly, when the 2019 fuel efficiency statistics came out, Southwest came in second place across all major U.S. airlines.[18] This efficiency strategy also delivered low greenhouse gas emissions per passenger mile, once again providing a win for investors concerned about climate change as well as airline profitability.

Unfortunately, none of today's ESG data providers offer metrics that get into this sort of nitty-gritty assessment of corporate sustainability strategies and implementation plans. Their ESG taxonomies largely emerged from the world of socially responsible investing (SRI)—and thus, their data frameworks largely reflect the base of issues (now broadening) that concerned *values*-oriented stakeholders of various types. These traditional SRI concerns are real and valid, but the metrics they spurred were never intended to drive insights on how a firm intends to create long-term *value* or competitive advantage through its sustainability efforts. Despite good work by groups such as the Sustainability Accounting Standards Board (SASB) and the Task Force on Climate-related Financial Disclosures (TCFD), who have tried to sharpen the focus on *material* ESG factors in various industry sectors, little has been done to develop the required deep-dive studies. A business strategy-focused analysis that looks at companies from the *inside out* rather than *outside in* would do much more to help investors understand a company's sustainability trajectory and identify what is *uniquely important* to each firm based on their value creation strategy or business model.

To be clear, what we are saying is not all that revolutionary. Traditional investment analysts across almost all business sectors carefully study corporate strategy, and they look for evidence both in a company's required financial reporting as well as in other non-financial metrics that companies provide to determine if the strategic vision looks promising. They further assess whether the company has the capacity and competitive posture to execute its strategy—and thus deliver marketplace success. Ultimately, they look for indicators from the company that they are on track to generate the promised business results. Consider in this regard, Paul Polman's 2010 commitment, as the newly appointed Unilever CEO, to reset the company's business strategy with a goal of doubling revenue while halving environmental impact by riding the sustainability megatrend and restructuring the company's products and supply chains to produce better outcomes for people and planet.[19] In doing so, he prodded analysts in the fast-moving consumer goods sector to rethink how sustainability might drive *future* growth, productivity, and risk-avoidance in the sector. By 2016, Unilever was reporting that sustainable products were growing 50% faster than standard products and were responsible for 60% of Unilever's growth.[20] When presented in this context, mainstream analysts had a *need to know* about sustainability.

This kind of cause-and-effect exploration of corporate vision and strategy execution—and the related methods of mapping that logic to business results—will be familiar to mainstream investment analysts and advisors concerned about business performance and financial results. But this approach is fundamentally different from the sort of data that today's ESG taxonomy-based research offers. To be clear, these traditional metrics have and will continue to have an important role in benchmarking companies—especially if the methodological weaknesses that have been noted are corrected.

As the sustainability imperative gains strength in the coming decade—unleashing new pressures not just for decarbonization, but also for healthy foods, living wages, management diversity, corporate transparency, privacy protection, and more—many investors will want better ESG analytics. In particular, they will want help in identifying those companies whose pathway to future competitive advantage, like Unilever's, derives from successful execution of a business model transition that promises to deliver value based on the suite of sustainability drivers we have identified. In this regard, strategy-focused analytics offer important new ground to plow for those hoping to gain actionable insights on the power of the sustainability imperative to drive business and investing results.

We don't mean to suggest that developing these new *value driver ESG analytics* will be easy. To the contrary, much more effort will need to be put into finding ways to gauge the *sustainability premiums* that accrue to companies who capitalize on the changes in the marketplace that the sustainability megatrends promise to impose. But a next generation of ESG metrics that highlight the marketplace winners and losers from the sustainability imperative would profoundly broaden interest in sustainable investing among the vast swath of mainstream investors and traditional investment advisors who remain skeptical about the signals today's ESG metrics provide.

Strategy-focused *deep dive* perspectives will be especially critical in certain industries and sectors and for certain issues. For example, a granular view of the decarbonization strategies of companies in carbon-intensive sectors (e.g., oil and gas, utilities, airlines, automakers, and heavy manufacturing) will be essential to investor decision-making in the years to come. In the oil and gas industry, for example, major players such as BP, Shell, Repsol, and Total as well as smaller companies, such as Sweden's Lundin Petroleum, have announced their intentions to become carbon neutral by 2050 and to transition toward being clean energy providers. But some of their competitors are clinging to their existing markets and business strategies—and seem committed to playing out their string as fossil fuel producers. This division of perspectives makes detailed data on *carbon exposure* and the risk of *stranded assets* (oil or gas reserves that will never be able to be extracted because of changing regulatory requirements and market economics) ever more critical and yet in short supply.[21]

Likewise, companies that do not keep up with fast-changing public expectations about racial equality, wage fairness, workplace conditions, and corporate integrity will likely find themselves facing ever more significant pushback in the marketplace and across society. And some may even lose their *social license to operate* as fitness giant CrossFit found out when CEO Greg Glassman's mishandling of the death of George Floyd and suggestions of a sexist working environment caused his company to implode. Moreover, as Todd Cort's chapter in this volume spells out with reference to other recent headline-grabbing corporate missteps, such as the Deepwater Horizon oil spill and the VW "dieselgate" scandal, typical ESG metrics are too high-level and lack insight into management practices to spot looming sustainability failures before they erupt.

Felix Preston, Director of Sustainability Insights at Generation Investment Management, observed in a recent white paper: "today's ESG data has real limitations. The risk is that it puts the spotlight on what is available, rather than what is most important."[22] We agree. Too many sustainability-minded investors are relying on poor quality ESG data and metrics that miss the most important dimensions of sustainability because that is what the data marketplace provides. We share Preston's conclusions that what investors really need is "rich contextual information" and that meeting the demands of the sustainability imperative depends on "a future where high-quality companies are aligned with planetary and societal needs." All of this will require a more refined *next-generation* approach to ESG metrics.

Next-Generation Sustainability Metrics

We see the evolution of sustainability metrics moving forward on two fronts. First, looking from the *outside in* at how companies perform on key issues of interest to sustainability-oriented investors and other stakeholders requires refined ESG metrics that will allow clear differentiation among companies with regard to their prospects for success in a future that has been redefined by the sustainability-driven changes sweeping across society. We note that the requirements for business success going forward are evolving rapidly based on changing government regulations, health implications, marketplace calculations, and—most importantly—public expectations. This new framework of ESG metrics will therefore need to track issues across the sustainability agenda, including whether companies are helping to deliver a decarbonized economy and progress on other environmental issues such as air and water pollution, waste handling, chemical exposures, and natural resource management. But they will also need to provide a clearer picture, based on methodologically strengthened ESG metrics, of which entities are leading (or lagging) on a range of other concerns including: (1) diversity within the management ranks as well as the board of directors, and promotion of an inclusive workplace; (2) investments in human capital and training; (3) pay equity and worker satisfaction; (4) adherence to labor and human rights standards; (5) customer experience including issues of privacy, trust, and responsiveness; (6) business integrity and transparency; and (7) corporate governance best practices.

Second, looking from the *inside out*, there will be increasing interest in firm-level, strategy-focused tools and metrics where traditional thinking about ESG scores almost gets turned upside down. The key question will not be the ESG score, but how sustainability-related drivers affect business strategy and thus translate into value based on gains in growth, productivity, and risk management.

Simply put, if mainstream investors are to come to more of a causal (rather than correlational) understanding of the relationship between sustainability and business performance—something that is a prerequisite for broad-scale adoption of sustainable investing strategies—then the data provided by the sustainability experts must fit both the financial and mental models utilized by mainstream business analysts. Those models are rooted in core business strategy as evidenced by the attention paid to factors such as value creation and cash flow. The widely used methods for modeling business prospects, such as Porter's *activity map* or the Kaplan-Norton *strategy map*,[23] offer the foundation on which a whole new approach to assessing corporate sustainability leadership might be constructed.

In sum, investors increasingly need a *wide view* of the business landscape that gives a sharper picture of ESG performance across companies traded on stock exchanges (and ultimately privately held corporations, as well) built on refined metrics of high-quality and clear comparability. But investors would also benefit from a *deep view* of the capacity of companies to design and deliver strategies that, not just address, but capitalize on the sustainability-driven challenges and opportunities of a changing business world.

Notes

1. Leiserowitz, A. et al. (2020, May 19). Climate Change in the American Mind: April 2020. *Yale Program on Climate Change Communications.*
2. Lubin, D. A., & Esty, D. C. (2010, May 1). The Sustainability Imperative. *Harvard Business Review.* Retrieved from https://hbr.org/2010/05/the-sustainab ility-imperative.
3. See, for example, the Open Back Better Act of 2020, 116th Cong., 2nd Sess. (2020) (introduced July 1, 2020 by Senator Tina Smith of Minnesota); The Royal Household. (2020, June 3). The Prince of Wales's Sustainable Markets Initiative launches #TheGreatReset. *Royal.uk.* Retrieved from https://www. royal.uk/prince-wales%E2%80%99s-sustainable-markets-initiative-launches-thegreatreset; United Nations Department of Global Communications (2020, April 22). Climate Change and Covid-19: UN Urges Nations to 'Recover Better.' *United Nations.* Retrieved from https://www.un.org/en/un-corona virus-communications-team/un-urges-countries-%E2%80%98build-back-bet ter%E2%80%99; Whiting, K. (2020, June 3). How the World Can 'Reset' Itself After COVID-19—According to These Experts. *World Economic Forum.* Retrieved from https://www.weforum.org/agenda/2020/06/covid19-great-reset-gita-gopinath-jennifer-morgan-sharan-burrow-climate/.
4. Such commitments include Amazon's $2 billion climate fund, *see* Amazon. (2020, June). Climate Pledge Fund. *Amazon.* Retrieved from https://sustainab ility.aboutamazon.com/about/climate-pledge-fund; Mattioli, D. (2020, June 23). Amazon to Launch $2 Billion Venture Capital Fund to Invest in Clean Energy. *The Wall Street Journal.* Retrieved from https://www.wsj.com/articles/ amazon-to-launch-2-billion-venture-capital-fund-to-invest-in-clean-energy-115

92910001?mod=hp_lead_pos7; Microsoft's sweeping carbon offset commitment, *see* Smith, B. (2020, January 16). Microsoft Will Be Carbon Negative by 2030. *Official Microsoft Blog*. Retrieved from https://blogs.microsoft.com/blog/2020/01/16/microsoft-will-be-carbon-negative-by-2030/; Google's $175 million commitment to racial equity, see Pichai, S. (2020, June 17). A Message from Our CEO: Our Commitments to Racial Equity. *Google—The Keyword*. Retrieved from https://www.blog.google/inside-google/company-announcem ents/commitments-racial-equity/; Merck's $20 million contribution to address health inequalities and Covid-19 relief, see Merck & Co., Inc. (2020, May 5). Merck Commits Additional $10 million to COVID-19 Relief Efforts to Help Disparately Impacted Patients and Communities. *Merck*. Retrieved from https://investors.merck.com/news/press-release-details/2020/Merck-Com mits-Additional-10-Million-to-COVID-19-Relief-Efforts-to-Help-Disparately-Impacted-Patients-and-Communities/default.aspx; and Walmart's decision to commit $100 million to create a new center on racial equity, see McMillon, D. (2020, June 12). Advancing our Work on Racial Equity. *Walmart*. Retrieved from https://corporate.walmart.com/newsroom/2020/06/12/advancing-our-work-on-racial-equity; Repko, M. (2020, June 5). 'What We See Is a Moment Here': Walmart CEO Doug McMillon Said Nation's Corporate Leaders Must Advance Racial Equity. *CNBC*. Retrieved from https://www.cnbc.com/2020/06/05/walmart-ceo-doug-mcmillon-said-ceos-must-advance-racial-equality. html; McMillon, D. (2020, June 5). Making a Difference in Racial Equity: Walmart CEO Doug McMillon's Full Remarks. *Walmart*. Retrieved from https://corporate.walmart.com/equity.

5. Makower, J. (2020, June 29). Trumping ESG. *GreenBiz*. Retrieved from https://www.greenbiz.com/newsletter/trumping-esg; Benjamin, J. (2020, April 19). As Pandemic Rages on, ESG Funds Shine Brightly. *Investment News*.

6. Esty, D., & Cort, T. (2017) Corporate Sustainability Metrics: What Investors Need and Don't Get. *The Journal of Environmental Investing, 8*(1). Retrieved from http://www.thejei.com/wp-content/uploads/2017/11/Journal-of-Environmental-Investing-8-No.-1.rev_-1.pdf.

7. Lubin, D., & Esty, D. (2014, June 17). Bridging the Sustainability Gap. *MIT Sloan Management Review*.

8. Preston, F. (2019, December 5). The Future of ESG Data. *Generation Investment Management*. Retrieved from https://www.generationim.com/research-cen tre/insights/the-future-of-esg-data/.

9. We note that others will have different views of what the sustainability megatrend might encompass. See, for example, Johan Rockstrom's "Planetary Boundaries." Rockstrom, J. et al. (2009). A Safe Operating Space for Humanity. *Nature*; Garrison Institute. (2020). *Pathways to Planetary Health*. Retrieved from https://www.garrisoninstitute.org/programs-retreats/programs/pathways-to-planetary-health/.

10. Business Roundtable. (2019). Statement on the Purpose of a Corporation. *Business Roundtable*. Retrieved from https://opportunity.businessroundtable.org/our

commitment/; Henderson, R. (2020). *Reimagining Capitalism in a World on Fire*. New York: Public Affairs.

11. Esty, D. C. (2019). Red Lights to Green Lights: Toward an Innovation-Oriented Sustainability Strategy. In D. C. Esty (Ed.), *A Better Planet: Forty Big Ideas for a Sustainable Future*. New Haven: Yale University Press.

12. Zuboff, S. (2019). *The Age of Surveillance Capitalism: The Fight for a Human Future at the New Frontier of Power*. New York: Public Affairs; Heineman, B. (2008). *High Performance with High Integrity*. Boston: Harvard Business School Publishing; Tapscott, D., & Ticoll, D. (2003). *The Naked Corporation: How the Age of Transparency will Revolutionize Business*. New York: Free Press.

13. Lubin, D. A., & Esty, D. C. (2010, May 1). The Sustainability Imperative. *Harvard Business Review*.

14. Lubin, D. A., & Esty, D. C. (2014, June 17). Bridging the Sustainability Gap. *MIT Sloan Management Review*. Retrieved from https://sloanreview.mit.edu/article/bridging-the-sustainability-gap/.

15. Lubin, D., & Krosinsky, C. (2013). The Value Driver Model: A Tool for Communicating the Business Value of Sustainability. *Principles for Responsible Investing* and *UN Global Compact*.

16. Kaplan, R. S., & Norton, D. P. (2004). *Strategy Maps: Converting Intangible Assets in Tangible Outcomes*. Cambridge: Harvard Business School Publishing.

17. Porter, M. E. (1996). What Is Strategy? *Harvard Business Review*. Retrieved from https://hbr.org/1996/11/what-is-strategy.

18. Zheng, X. S., Graver, B., & Rutherford, D. (2019). *U.S. Domestic Airline Fuel Efficiency Ranking, 2017–2018*. International Council on Clean Transportation.

19. Unilever. (2011). *Unilever Sustainable Living Plan*. https://www.unilever.com/Images/unilever-sustainable-living-plan_tcm244-409855_en.pdf.

20. Unilever. (n.d.). *Making Purpose Pay: Inspiring Sustainable Living*. https://www.unilever.com/Images/making-purpose-pay-inspiring-sustainable-living-170515_tcm244-506419_en.pdf.

21. The Rocky Mountain Institute is developing this sort of granular analysis. Dyson, M., Engel, A., & Farbes, J. (2018). *The Economics of Clean Energy Portfolios: How Renewable and Distributed Energy Resources Are Outcompeting and Can Strand Investment in Natural Gas-Fired Generation*. Rocky Mountain Institute. Retrieved from https://rmi.org/wp-content/uploads/2018/06/RMI_Clean_Energy_Portfolios_2018.pdf.

22. Preston, F. (2019, December 5). The Future of ESG Data. *Generation Investment Management*. Retrieved from https://www.generationim.com/research-centre/insights/the-future-of-esg-data/.

23. See Porter (1996) at note 17 above, and Norton and Kaplan (2004) at note 16 above.

Part III

ESG Financial Products

8

Financing the Transition to Green Infrastructure

Ella Warshauer and Cary Krosinsky

Abstract Transitioning to a low-carbon economy and building a sustainable twenty-first-century society will mean remaking the infrastructure of our communities, industrial sector, transportation networks, and above all our energy system—which will require vast amounts of capital. This transition is a significant opportunity to lay the foundations for a sustainable future. There remains, however, an infrastructure finance gap as the demand for funding far exceeds the supply of capital required. Mobilizing the requisite $90 trillion in capital by 2030 will not be easy, but that is what it will take if investors are serious about achieving the carbon reductions necessary to prevent severe climate change and to mitigate irreversible environmental and economic damage.

Keywords Infrastructure · Financing gap · Green infrastructure · Sustainability · ESG negative exclusions · Shareholder advocacy · Fixed income · Property · Real assets · Green bonds · Sustainability framework · Greenwashing · Public equity · Exchange traded funds · ETFs · Public-private partnerships

E. Warshauer (✉)
Powerhouse Ventures, Oakland, CA, USA

C. Krosinsky
Yale and Brown University, Oakland, CA, USA

© The Author(s) 2020
D. C. Esty and T. Cort (eds.), *Values at Work*,
https://doi.org/10.1007/978-3-030-55613-6_8

111

Climate change poses an existential threat to the world and, as a number of the chapters in this book explain, creates increased uncertainty and potential risk for both companies and investors. But the need to build a clean energy future and respond to the *sustainability imperative* more broadly offers opportunities as well as exposure.[1] In brief, transitioning to a low-carbon economy and building a sustainable twenty-first-century society will mean remaking the infrastructure of our communities, industrial sector, transportation networks, and above all, our energy system—which will require vast amounts of capital. But this transition must be seen as a significant opportunity to lay the foundations for a sustainable future. There remains, however, an infrastructure finance gap as the demand for funding far exceeds the supply of capital for financing the required investments.

Infrastructure is critical to the successful daily functioning of every modern society, yet the stock of global infrastructure sorely needs to be revitalized. Hundreds of bridges in Italy are at risk of collapse—as Genoa learned all too painfully in 2018—and aging power plants in Europe suffer from inefficiencies. In 2017, the United States scored a D + on its Infrastructure Report Card, released by the American Society of Civil Engineers.[2] A renewed focus on infrastructure investing, especially as societies seek to rebound from the COVID-19 pandemic, would meaningfully contribute to economic growth and job creation. More importantly, achieving a sustainable future depends on a well-planned transformation of the world's infrastructure stock.

This chapter outlines the magnitude of the infrastructure finance gap, defines the role of green infrastructure, and highlights relevant investment opportunities across asset classes, especially treasury bonds, physical "real" assets, property, and private equity.

Infrastructure and Climate Risk

As the Intergovernmental Panel on Climate Change's (IPCC) Special 2018 Report on Global Warming shows, 2 degrees Celsius of global warming would cause significant environmental and economic damage across much of the world. The report anticipates direct physical risks from warming, such as increased intensity and frequency of extreme weather events, as well as substantial dislocation to agriculture, manufacturing, and other production activities, which would have far-reaching implications for businesses, governments, and households.

With the scope of the potential impacts in mind, investors increasingly want to factor climate change risk into their investment decisions. Central banks, financial institutions, and the public sector—through organizations such as the Network for Greening the Financial System and the Task Force on Climate-Related Financial Disclosure—all acknowledge the risks posed by a warming world, as well as all of the follow-on effects this promises in terms of sea level rise, increased intensity and frequency of hurricanes, changed rainfall patterns, ecosystem impacts, and new vectors of diseases that could be potential pandemics.

To avoid irreversible damage, global greenhouse gas emissions must be cut dramatically by 2030 and fall to net-zero by 2050, a task which will require broad global action. Delivering on this scale of decarbonization requires a near-complete turnover of the world's existing infrastructure stock—including all of the core energy, transportation, building, waste, land-use, and water systems that support the day-to-day functioning of the global population. Transitioning to green infrastructure—such as renewable power generation, high-efficiency buildings, electrified vehicles, and circular solutions for waste management and industry—will be necessary.

What Is Green Infrastructure?

Traditional infrastructure investing is thought of as *duration* infrastructure, meaning long-duration projects. Examples of traditional infrastructure projects would include highways, railways, ports, energy, water, airports, and telecommunications infrastructure. Infrastructure is considered one of the least digitally transformed sectors in the global economy. Moreover, much of today's infrastructure is resource-intensive and built without regard for long-term sustainability.

Green infrastructure centers on projects that optimize and incorporate environmental, social, and governance (ESG) aspects into project planning, construction, and operations. Examples can include electrification and clean transportation systems, the incorporation of green building and design principles into the built environment, and the development of renewable energy technologies, circular solutions for sustainable waste management, and permeable surfaces for stormwater management.

In contrast to traditional infrastructure, green infrastructure projects are designed to be resilient in the face of natural hazards, achieve low-carbon and environmentally sustainable outcomes, and advance the UN Sustainable Development Goals[3]—utilizing efficient materials, technologies, and systems throughout the design and build process. Green infrastructure practices generate positive spillovers for society with benefits that might contribute to reduced inequality, improved health outcomes, and expanded economic development.

The decisions we make in rebuilding the world's infrastructure today will have long-term implications for generations to come. To transition to a resilient, low-carbon economy and ensure the world's infrastructure stock can meet the needs of the growing population, we must employ green infrastructure practices that can withstand the physical risks posed by climate change and avoid "locking in" societies with carbon-intensive—or otherwise unsustainable—projects.

The Infrastructure Investment Gap

The current value of the world's infrastructure stock is approximately $50 trillion, yet global demand for infrastructure is far greater, requiring an estimated $90 trillion of investment from 2015 to 2030. This magnitude of infrastructure investment demand is tantamount to *literally* rebuilding our world.[4]

While funding for infrastructure projects reached a peak in 2017, closing the infrastructure investment gap, which is necessary to remain below the 2-degree warming target, will require investment at three to five times current levels, or as much as $6 trillion per year. Energy, transport, and water and waste management projects represent the most significant infrastructure needs to be met within the ongoing financing gap, accounting for 43%, 29%, and 21% of projects, respectively.

Closing this finance gap and building green infrastructure requires investors to approach infrastructure investing from a fundamentally different perspective. Given the rapid aging of the world's current infrastructure stock, the magnitude of the financing gap, and the resulting magnitude of the investment opportunity, investors can no longer afford to ignore traditional duration infrastructure's environmental impacts.

Fortunately, investors have finally begun to realize that failure to respond seriously to climate change will have devastating consequences for economic output and global prosperity. As a result, sustainable investing strategies are being incrementally adopted across the largest asset classes, primarily in fixed income and real assets. Total global assets allocated to sustainable strategies have grown from $13.3 trillion in 2012 to $30.7 trillion in 2018. In the United States, sustainable investing strategies grew 38% to $12 trillion from 2012 to 2018, with asset management firms holding $11.6 trillion. While these numbers suggest rapid growth, the truth is that the ESG focus is often limited to mere screening by *negative exclusions*—where companies involved in selling tobacco or guns are simply excluded from portfolios. Another broad set of sustainability strategies narrowly reflect some form of shareholder advocacy, which can involve little to no direct investment in climate change action or other real sustainability solutions.[5] Still, trillions remain necessary for adequate levels of investment to be achieved in green infrastructure, and investors, governments, financial institutions, and policymakers have their work cut out for them.[6]

The Value of Everything

From an investor's perspective, sustainable infrastructure opportunities exist across all asset classes—offering a wide variety of investment vehicles that will advance the transition to green infrastructure. Fixed income, property, and real assets are some of the largest asset classes that require transformation.

In the sections that follow, we review the current state of capital deployed across asset classes, outline strengths and weaknesses of existing mechanisms as related to intended sustainability-related outcomes, and discuss creative financing options that investors can leverage to close the existing financing gap. We will start by examining the state of global asset classes (Table 8.1).

In late 2015, the United Nations Environment Programme calculated the global value of assets to be $450 trillion.[7] The largest asset classes were fixed income—primarily treasury bonds—at 22%, and real estate or property, at 21%. In 2020, the value of everything is believed to have increased to approximately $500 trillion, though the 2020 pandemic likely affected these valuations, driven primarily by increases in fixed income now estimated at $116 trillion, and the rise in the value of public equity investment. For duration infrastructure and real assets, increased investment has

Table 8.1 The value of everything (United Nations Environment Program. (2015). *The Value of Everything.*)

$450 trillion USD, as of 2015	
Asset class	%
Fixed income	22
Property	21
Cash	17
Public equity	15
State-owned enterprises	8
Infrastructure	8
Derivatives	6
Real assets	2
Privately owned companies	1

been balanced out by aging and depreciation, leaving these asset classes close to previous levels. Property value has also likely decreased slightly due to these factors.

Within the value of everything, the global value of infrastructure specifically has been estimated at $20 trillion,[8] and private investment accounts for about half of total infrastructure spending, $1 to 1.5 trillion per year.[9] The value of treasury bonds and other fixed income instruments represents the overall amount of money lent out in the economy, and is estimated to be around $116 trillion, making fixed income the largest asset class.[10] Regarding infrastructure in particular, debt financing represents on average 70% of total allocated capital.[11]

Green Bonds

Green bonds represent one specific mechanism for financing projects with an environmental focus that has been successful in scaling-up to help facilitate the transition to a sustainable future. 2019 saw approximately $200 billion of green bonds loaned out, or $1 trillion cumulatively since 2007,[12] on pace to hit $1 trillion per year by 2023,[13] according to Sean Kidney of the Climate Bonds Initiative.[14]

Green bonds are mostly used to raise funds for climate change solutions and other projects that are intended to have some positive environmental aspect. Green bonds particularly stand out as an appealing option for pension funds, insurance companies, and investment institutions that seek long-term, steady cash flows and low correlations with other assets, aspects which also appeal to infrastructure investors. Green bonds must undergo a certification process using a sustainability framework, such as that of the Climate Bonds Initiative. Later, bonds may undergo a second opinion, which is an independent assessment of the bond's sustainability framework and is used to confirm issuers' internal procedures. Companies such as Cicero and Sustainalytics offer second opinion assessments.

Nobina's Green Bond for Clean Transport

Nobina, the Nordic region's largest public transport company, is a valuable case study of a successfully executed green bond. This bond qualified as "dark green," the most highly ranked category in Cicero's Shades of Green second opinion framework.

Dark green is allocated to projects and solutions that correspond to the long-term vision of a low-carbon and climate resilient future. Fossil-fuel technologies that lock-in long term technologies do not qualify for financing. Ideally, exposure to traditional and physical climate risk is considered or mitigated. Example projects include wind energy projects with a strong governance structure that integrates environmental concerns.[15]

This qualification means that the project is on track to support a low-carbon and climate-resilient future. Bond proceeds are used to finance clean transport, such as vehicles powered by biofuels and charging infrastructure for electric buses.[16] Nobina successfully and transparently reports on the use of proceeds, which were tracked and credited to a separate, designated account. The main pitfall of this transportation infrastructure project is that it still contributes to greenhouse gas emissions through the use of biofuels, and some analysts fear that Nobina will lock in technologies that are not on the most aggressive track to mitigate climate change.[17]

Drawbacks of Green Bonds

While green bonds have been well-received by the investment community, seeing growth in both issuance and average deal size, they do have some drawbacks. Green bonds must issue impact reports, which provide transparency into the bond's use of proceeds, as well as other relevant impact metrics. There can be a significant learning curve for investors not experienced in green bond issuance regarding implementing sufficient auditing and accounting processes.[18] This inefficiency can result in investors paying a premium to acquire green bonds in the secondary market—Barclays's research indicates that this premium may be as much as 20%.[19]

In addition, there are concerns around *greenwashing*, meaning some projects that are not considered *green* have nonetheless been financed with green bonds. For example, in 2015, China included "clean coal" in a central bank list of technologies eligible for green bonds.[20] If investors are serious about staying on track to remain below a 2-degree scenario, green bonds should only be issued if projects truly contribute to an ongoing transition to a low-carbon future.

Green bond investments in infrastructure that meaningfully contributes to the low-carbon transition will be critical to society's ability to meet the goals of the 2015 Paris Climate Change Agreement. But achieving the necessary scale will require: (a) ramping up allocated assets to $1 trillion or more per year, (b) identification of projects that provide true additionality from a greenhouse gas emissions perspective, perhaps using Project Drawdown or similar analysis, (c) project-level analysis that ties the intentionality of the carbon reduction to project design, (d) ongoing verification that projects have indeed been reducing carbon over time, and (e) ways to do all of the above without adding significant costs so that green bonds become financially unattractive.

Governments can also play a useful role in the development of sovereign wealth funds that can allocate such capital, providing financial incentives for green issuance and establishing appropriate policies that encourage the necessary flow of capital in this important direction.

Property

Property is the second-largest global asset class. At $95 trillion, it represents the value of peoples' homes and managed portfolios of real estate.[21] The energy footprint of property is significant, and thus, this asset class represents a large opportunity for investors interested in accelerating the transition to a low-carbon economy.

Property offers investors the chance to invest in sustainable design, reductions in energy- and water-use, greenhouse gas emissions reduction, air quality improvement, and natural resource protection. These improvements are integral components of green infrastructure projects, and there is significant overlap between industry, property, and infrastructure more generally.[22] An increase in energy-efficient buildings is considered essential for greening U.S. infrastructure.[23]

While certifications such as Leadership in Energy and Environmental Design (LEED) and Building Research Establishment Environmental Assessment Method (BREEAM) provide useful guidelines for green buildings, and energy benchmarking policy provides baseline data for evaluating performance, these approaches have not been sufficient for reducing emissions and increasing efficiencies.[24] Studies demonstrate LEED buildings are oftentimes actually less efficient than their counterparts, more expensive to build due to increased costs for consultants, and therefore more expensive to occupy.[25] The process for becoming LEED certified has also been touted as superficial and gimmicky, especially in comparison with the Environmental Protection Agency's Energy Star program where buildings must submit actual utility bills before receiving certification.[26]

Overall, the buildings sector, which accounts for 36% of final energy use and 39% of energy-related carbon emissions, is not keeping pace with population growth and increased demand for energy services.[27] If a more ambitious effort is not made to improve energy efficiency in buildings, final energy demand from the sector could increase 30% by 2060, which would falter off track for a 2-degree warming scenario.[28]

Other mechanisms that investors can draw upon but which are not explained in detail in this chapter include: Property Assessed Clean Energy (PACE) financing[29] and Energy Performance Contracting (EPC).[30]

Recommendations

While the use of the Leadership in Energy and Environmental Design and Building Research Establishment Environmental Assessment Method program is generally well accepted and creates useful certifications, projects using these standards do not always lead to carbon reductions. Better standards are needed to demonstrate quantifiable carbon emissions reductions. Retrofitting programs for older buildings could also help generate meaningful jobs in the age of the COVID-19 pandemic, especially given International Energy Agency scenarios, which show efficiency is a key pillar for achieving emissions reductions more generally.

Public Equity

The third-largest asset class is public equity, most recently estimated at $85 trillion.[31] Public equity has low transaction costs and requires less due diligence compared to other forms of direct investment in green infrastructure.[32] It therefore plays an important role in helping to scale up the amount of capital deployed by involving investors who might be interested in infrastructure, but not direct investment in sustainability projects. Engaging new investors is positive but an important question remains: Can investing in an existing company really have a significant impact on transitioning to a low-carbon economy?

Exchange Traded Funds

Exchange traded funds (ETFs) offer another way for investors to allocate capital toward green infrastructure. ETFs are a package of securities that track an index and trade on an exchange. There are a number of "green" ETFs currently on the market, mostly branded as clean energy or broad

ESG funds. Examples of sustainability-oriented ETFs include: iShares Clean Energy (ICLN), First Trust Nasdaq Clean Edge Green Energy (QCLN), and First Trust's NASDAQ Clean Edge Smart Grid Infrastructure Index (NYSE: GRID). Investors can also promote a sustainable infrastructure marketplace through green bond ETFs, such as the iShares Global Green Bond ETF. Currently, the market for green infrastructure ETFs is small, and traditional investors see these plays as niche when it comes to portfolio strategy. Still, these early players play an important role in growing the green infrastructure ETF market, and will benefit from a strong track record when the market is big enough to support these investments at scale.

Public equity is one of the largest segments of the value of all tradeable assets. The expansion of options which connect directly to green infrastructure would unlock a large, untapped pool of investment assets which could be critical for bridging the green infrastructure investment gap, especially in the United States.

Duration Infrastructure and Real Assets

Infrastructure is widely considered its own asset class and traditionally includes projects with 35- to 100-year timelines, such as highways, railways, ports, water, airports, and telecommunications systems. The long-term investment time frame of infrastructure projects is often characterized by public-private partnerships. Real assets, such as land, commodities, and equipment, sit outside of public companies in private hands, and are inextricably tied to infrastructure. Real assets are traditionally thought of as a separate asset class and are estimated at $10 trillion, which is 2% of global assets under management.[33]

Pensions and sovereign wealth funds, particularly in the United States, are some of the largest investors in infrastructure and can play a larger role in transitioning to green infrastructure. They represent $10.9 trillion and $6.3 trillion in assets under management, respectively, have long time horizons and can afford to take longer-term risks.[34] These funds, however, have a mandate to diversify assets to reach target rates of return, and infrastructure, on average, generates lower returns on investment than other sectors, requires time-intensive due diligence, and comes with additional transaction costs. As a result, there are concerns that significant allocations to infrastructure cannot always help asset owners achieve targeted annual return goals. This remains a significant barrier in scaling-up private investment in infrastructure.

In addition, infrastructure project development can be complicated by the involvement of many stakeholders and time-intensive due diligence. As a result of the complexities involved in developing infrastructure, projects often require different types of financing at various stages of a project's life cycle, and it is common for investors to implement public-private partnerships to scale infrastructure projects. Overall, the variety of factors involved in infrastructure project development can be challenging for newer investors to overcome and represent additional barriers to ramping up allocations to infrastructure.

If investors are serious about enabling green infrastructure development, the private sector must avoid "locking in" to traditional infrastructure, and prioritize investing in green infrastructure. While developing expertise in green infrastructure projects may present new investors with a steep learning curve, the magnitude of the financing gap, the significance of the investment opportunity, and the positive impacts of green infrastructure for transitioning to a low-carbon economy are too large to ignore.

Further Considerations

Increasing rates of capital flowing into sustainable investment strategies indicate that the industry is starting to mature, but a much larger flow of assets toward green infrastructure will be necessary to achieve desired societal outcomes.

Challenges remain in ramping up investments from investors who might not traditionally invest in infrastructure, and in the provision of clear infrastructure investment frameworks, processes, incentives, and knowledge to help mobilize capital more quickly, efficiently, and effectively. Closing the financing gap means changing the way investors approach infrastructure as an asset class. Private equity poses one possible solution.

Private equity investors are often majority or sole owners of large infrastructure operations, such as airports and seaports, and could become more deeply involved with owning and transitioning utilities, ensuring pipeline safety, helping to minimize spills, or by overseeing cleanup operations. Private equity has already started to mobilize capital to finance clean energy. According to Bloomberg New Energy Finance, of the $363 billion invested globally in clean energy, private equity and venture capital committed $10.5 billion in 2019, up 6% from 2018, and the highest amount since 2010.[35] Private equity investors can also be less constrained than traditional asset owners and often take a more direct approach to their investments. This is important for investing in green infrastructure projects, which often involve more due diligence and investment expertise when compared to investing in traditional infrastructure.[36] The challenge and opportunity is to match dry powder—or uninvested commitments of capital—with investable projects.

Public-private partnerships and green banks may also help investors more easily identify and mobilize capital to investable green infrastructure projects. Public entities can be more transparent with their infrastructure pipelines to provide the private sector with a clearer understanding of infrastructure needs and allowing for efficient mobilization of capital. Green banks, which are publicly capitalized entities designed specifically to mobilize private capital to low-carbon, climate-resilient infrastructure, can also play a larger role. Green banks provide investors with market expertise and can assist in implementing innovative transaction structures.

Private equity, public-private partnerships, and green banks can make a significant impact on the transition. However, more drastic measures may be needed to mobilize the requisite amount of capital for sustainable infrastructure, particularly given the potential long-term impacts of the COVID-19 pandemic on the need to generate good jobs and related income for working families. In addition to focusing on catalyzing private equity investments, public-private partnerships, and developing green banks, a green stimulus[37] may help jumpstart a shift in investor behavior and catalyze green infrastructure investment to the necessary levels.

Such a green stimulus could coincide with the creation of a new asset class exclusively dedicated to green infrastructure. Projects such as solar and wind farms, battery storage, electric vehicle charging, waste-to-energy, and other initiatives that correspond with the vision of a low-carbon future would fall into this new asset class.

Mobilizing $90 trillion in capital by 2030 will not be easy, but that is what it will take if investors are serious about achieving the carbon reductions necessary to ensure no more than 2 degrees of global temperature rise, and to mitigate irreversible environmental and economic damage.

Further research is needed to develop and define a new asset class for green infrastructure or an adequate level of green stimulus. These measures, coupled with effective existing, scalable solutions described in this chapter, could be the key to changing investors' perspectives on green infrastructure and catalyzing change.

Notes

1. Lubin, D., & Esty, D. (2010, May). The Sustainability Imperative. *Harvard Business Review*; Esty, D., & Winston, A. (2009). Green to Gold: How Smart Companies Use Environmental Strategy to Innovate, Create Value, and Build Competitive Advantage.
2. Ironcore. (2020). *ASCE's 2017 American Infrastructure Report Card: GPA: D.* Retrieved from https://www.infrastructurereportcard.org/.
3. United Nations. (n.d.). About the Sustainable Development Goals—United Nations Sustainable Development. Retrieved from https://www.un.org/sustainabledevelopment/sustainable-development-goals/.
4. Bielenberg, A. (n.d.). The Next Generation of Infrastructure. Retrieved from https://www.mckinsey.com/industries/capital-projects-and-infrastructure/our-insights/next-generation-of-infrastructure.
5. 2018 Global Sustainable Investment Review. (2018). Retrieved from http://www.gsi-alliance.org/wp-content/uploads/2019/03/GSIR_Review2018.3.28.pdf.
6. Ghosh, I. (2020, February 5). Visualizing the Global Rise of Sustainable Investing. Retrieved from https://www.visualcapitalist.com/rise-of-sustainable-investing/.
7. Krosinsky C. (2015). *The Value of Everything.* United Nations Environmental Programme.
8. Nicolaisen, J., & Yngve S. (2015, March 25). Government Pension Fund Global—Investments in Infrastructure. *Norges Bank.*
9. Bielenberg, A. et al. (2016, January). Financing Change: How to Mobilize Private-Sector Financing for Sustainable Infrastructure. *McKinsey.*
10. J. P. Morgan Asset Management, "60% Of the Global Bond Market Is Outside the U.S., But Only 7% of an Average Portfolio Is Allocated to Fixed Income Internationally. Something Doesn't Add up: Https://T.co/b9sJiTLOyC," Twitter (Twitter, July 22, 2019), https://twitter.com/i/status/1153287770451918849.
11. Bielenberg, A. et al. (2016, January). Financing Change: How to Mobilize Private- Sector Financing for Sustainable Infrastructure. *McKinsey.*
12. Fatin, L. (2019, November 30). Green Issuance Surpasses $ 100 Billion Mark for 2019: First Time Milestone Is Reached in First Half: EU TEG to Open 2020 Pathways Towards $1trillion. *Climate Bonds Initiative.* Retrieved from https://www.climatebonds.net/2019/06/green-issuance-surpasses-100-billion-mark-2019-first-time-milestone-reached-first-half-eu.
13. Fatin, L. (2019, November 30). Green Issuance Surpasses $ 100 Billion mark for 2019: First Time Milestone Is Reached in First Half: EU TEG to Open 2020 Pathways Towards $1trillion. *Climate Bonds Initiative.* Retrieved from https://www.climatebonds.net/2019/06/green-issuance-surpasses-100-billion-mark-2019-first-time-milestone-reached-first-half-eu.

14. Filkova M., Frandon-Martinez, C., & Giorgi, A. (2018). Green Bonds, The State of the Market 2018. *Climate Bonds Initiative.*
15. Cicero. (n.d.) *Nobina Green Bond Second Opinion*
16. Cicero. (n.d.). *Nobina Green Bond Second Opinion* pp. 4.
17. Cicero. (n.d.) *Nobina Green Bond Second Opinion*
18. Filkova M., Frandon-Martinez, C., & Giorgi, A. (2018). Green Bonds, The State of the Market 2018. *Climate Bonds Initiative.*
19. Filkova M., Frandon-Martinez, C., & Giorgi, A. (2018). Green Bonds, The State of the Market 2018. *Climate Bonds Initiative.*
20. Stanway, D. (2019, March 21). China to Cut Coal from New Green Bond Standards: Sources. *Reuters.* Retrieved from https://www.reuters.com/article/us-china-bonds-environment/china-to-cut-coal-from-new-green-bond-standards-sources-idUSKCN1R20PP.
21. Krosinsky, C. (2015). *The Value of Everything.* United Nations Environmental Programme.
22. Jones, S., & Laquidara-Carr, D. (2018). *World Green Building Trends 2018.* Smart Market. pp. 1–80.
23. Majumder, B., & Bassett, L. (2019, January 29). Energy-Efficient Buildings Are Central to Modernizing U.S. Infrastructure. *Center for American Progress.* https://www.americanprogress.org/issues/green/news/2019/01/29/465520/energy-efficient-buildings-central-modernizing-u-s-infrastructure/.
24. Swearingen, A. (2014, May 1). LEED-Certified Buildings Are Often Less Energy-Efficient Than Uncertified Ones. *Forbes.* Retrieved from https://www.forbes.com/sites/realspin/2014/04/30/leed-certified-buildings-are-often-less-energy-efficient-than-uncertified-ones/#45e09d792554.
25. Swearingen, A. (2014, May 1). LEED-Certified Buildings Are Often Less Energy-Efficient Than Uncertified Ones. *Forbes.* Retrieved from https://www.forbes.com/sites/realspin/2014/04/30/leed-certified-buildings-are-often-less-energy-efficient-than-uncertified-ones/#45e09d792554.
26. Swearingen, A. (2014, May 1). LEED-Certified Buildings Are Often Less Energy-Efficient Than Uncertified Ones. *Forbes.* Retrieved from https://www.forbes.com/sites/realspin/2014/04/30/leed-certified-buildings-are-often-less-energy-efficient-than-uncertified-ones/#45e09d792554.
27. Abergel, T., Dean, B., &.Dulac, J. (2017). *UN Environment Global Status Report 2017.* UN Environmental Programme, p. 15.
28. Abergel, T., Dean, B., &Dulac, J. (2017). *UN Environment Global Status Report 2017.* UN Environmental Programme., p. 15.
29. Property Assessed Clean Energy Programs. (n.d.). Retrieved from https://www.energy.gov/eere/slsc/property-assessed-clean-energy-programs.
30. Energy Savings Performance Contracting. (n.d.). Retrieved from https://www.energy.gov/eere/slsc/energy-savings-performance-contracting.
31. Noland, D. (2019, December 28). Weekly Commentary: Just The Facts—December 27, 2019. Retrieved from https://seekingalpha.com/article/4314346-weekly-commentary-just-facts-december-27-2019.

32. Bielenberg, A. et al. (2016, January). Financing Change: How to Mobilize Private- Sector Financing for Sustainable Infrastructure. *McKinsey.*
33. Krosinsky, C. (2015). *The Value of Everything.* United Nations Environmental Programme.
34. Bielenberg, A. et al. (2016, January). Financing Change: How to Mobilize Private-Sector Financing for Sustainable Infrastructure. *McKinsey.*
35. Bloomberg New Energy Finance. (2020, January 16). Clean Energy Investment Trends, 2019. Retrieved from https://data.bloomberglp.com/professional/sites/24/BloombergNEF-Clean-Energy-Investment-Trends-2019.pdf.
36. Blackstone. (2019). *Seeking an Alternative: Understanding and Allocating to Alternative Investments*, pp. 1–20.
37. Robins, N. (2009, February 25). A Climate for Recovery: The Colour of Stimulus Goes Green. Retrieved from https://www.globaldashboard.org/wp-content/uploads/2009/HSBC_Green_New_Deal.pdf.

9

Private Equity and ESG Investing

Christina Alfonso-Ercan

Abstract This chapter explores the history and current state of sustainable investing within the private equity (PE) industry. In doing so, it presents evidence of a rising demand for a sustainable approach to private investments as well as the need for further development of environmental, social, and governance (ESG) data analysis to enhance investment decision-making in the private equity world. It argues that PE investors are strategically positioned to drive this shift toward investment standards that take account of ESG impacts and to create the sustainable investing tools needed to deliver greater transparency, better resource management, more careful risk mitigation, and higher returns than traditional investment strategies alone. Such a shift will require further market standardization, consolidation, and self-regulation among industry participants. The chapter concludes that, while implementation challenges remain, improved ESG tools, products, and services will ultimately yield a more efficient market as the private equity industry benefits from the unparalleled growth of sustainable investments.

Keywords Sustainable investing · ESG impacts · Venture capital · Private equity · ESG analysis · Transparency · Data standardization

C. Alfonso-Ercan (✉)
Madeira Global, New York, NY, USA
e-mail: christina@madeira-global.com

© The Author(s) 2020
D. C. Esty and T. Cort (eds.), *Values at Work*,
https://doi.org/10.1007/978-3-030-55613-6_9

Sustainable investing has now emerged at the forefront of the private equity industry. Its messaging and marketing, however, run optimistically ahead of its strategic execution to date. The primary factor driving its successful implementation has been the increasingly popular incorporation of environmental, social, and governance (ESG) tools and metrics in investment decision-making. This chapter argues that, when efficiently and strategically applied, ESG analysis can deliver greater transparency, smarter resource management and risk mitigation, more comprehensive diligence reviews, and higher returns than traditional investment strategies.[1] It further theorizes that a widespread transition to ESG-centric investing is imminent and suggests how and why private equity investors might accelerate this phenomenon.

History of Private Equity's Catalytic Role in Investing

Over the last 200 years, the global market and the United States in particular have been transformed by three instrumental eras: The Industrial Revolution, the Second Agricultural Revolution, and the Digital Revolution, each of which was spurred by private investment. Beginning with a handful of wealthy individuals and families, private enterprise investments proved successful in gaining wealth and influence while spurring industry paradigm shifts. J. P. Morgan, the Vanderbilts, and the Rockefellers are just a few noteworthy examples of such investors.[2]

Following World War II, a form of private equity, known as venture capital, emerged as a way to ignite public interest in private sector investments. In the following years, government regulation expanded private equity by creating further incentives for capital flows.[3] The rise of private equity financed technological breakthroughs that established the United States as a global political and economic powerhouse. These private capital flows helped foster the entrepreneurship required to compete with the Soviet Union throughout the Cold War, establish Silicon Valley as a global technology hub, and lay the foundation for the important role that private equity and venture capital funds play in today's investment arena. This context is extremely relevant to today's private equity industry, as it demonstrates the vital role that private investors and private sector investments have played in driving innovation and affecting change—outcomes that government funding, and even public enterprise, are simply not structured to achieve on the same timeline, if at all.

Private Equity as a Key Driver for Sustainable Investing

History has repeatedly demonstrated that private equity investors are extremely well-positioned to demand swift and responsible corporate action to address critical problems. Indeed, private equity investors help raise awareness about a range of global, social, and environmental challenges such as governance malpractice, unfair or unsafe working conditions, irresponsible resource management, and climate change. They are also responsible for targeting investments in businesses that successfully address these challenges. To differentiate sustainable private equity investments from traditional ones, the use of ESG tools—including assessments and metrics—has emerged as a key instrument. ESG tools enable more comprehensive diligence and analytical processes. These factors are largely qualitative in nature but have nonetheless proven to be vital to the bottom line.

A powerful recent example of the impact of ESG factors on financial performance is the devastating impact of the coronavirus outbreak on the global economy. The pandemic was unlikely to be factored into 2020 corporate financial models, and yet the relevance of ESG factors on investment performance could not be more evident. Retail spending figures have plummeted due to extended quarantines and corporate belt-tightening, leading to historic high rates of unemployment. The private equity industry, on the back of a decade-long period of successful fundraising and transaction volume, has proven critical to fighting the coronavirus by leveraging acquired technologies, having the capital and resources to quickly develop new ones, and supplying urgent equipment through creative solutions and increased manufacturing.[4]

The aim of sustainable investing is logical and simple: to deliver strong financial performance, as well as solutions to global challenges across the spectrum of ESG issues. Historical evidence suggests that private equity investors are a powerful driver of sustainability in providing much-needed risk capital and imposing ESG-driven investment decision-making and ESG performance tracking and reporting.[5] A 2018 publication issued by the International Finance Corporation confirms that private equity investors increasingly demonstrate value alignment with investment portfolios that aim to achieve sustainable development goals such as, "boost[ing] economic growth, improv[ing] living standards, and generat[ing] a variety of employment options."[6] Since traditional funding channels are often unavailable to smaller private businesses, private equity investors also have the ability to inject capital to fund attractive opportunities that foster innovation and sustainable solutions.[7]

Rising Prevalence of ESG in the Private Equity Industry

Research increasingly suggests private equity capital that seeks to achieve positive ESG performance is not at odds with positive financial returns. Perhaps the most comprehensive study on the relationship between ESG criteria and corporate financial performance ever performed was published in 2015 by Friede, Busch, and Bassen in the Journal of Sustainable Finance and Investment.[8] This meta-analysis of 60 studies found a nonnegative relationship between ESG and corporate financial performance in roughly 90% of cases, with the majority reporting a positive relationship. The authors also noted that a staggering $60 trillion of assets were managed by signatories of the UN Principle for Responsible Investing (UNPRI), indicating some degree of commitment to factoring ESG performance into portfolio choices.[9] The study concludes that "the orientation toward long-term responsible investing...requires a detailed and profound understanding of how to integrate ESG criteria into investment processes in order to harvest the full potential of value-enhancing ESG factors."[10]

In a review of the top ten global private equity firms (defined by total assets under management)—including The Blackstone Group, Neuberger Berman, Apollo Global Management, The Carlyle Group, KKR & Co., Bain Capital, and Vista Equity Partners—100% include a mention of ESG in their overall messaging or are in the process of incorporating ESG practices into their corporate strategy.[11] When Vista Founder Robert Smith was interviewed by Bloomberg in January 2020, he prioritized ESG issues, asking: "how do we get corporate leaders… to think about ESG as a forefront of what they do… and actually create sustainable returns, not just one-time returns?"[12] Further demonstrating ESG as a top priority, Smith announced that Vista would be purchasing 650,000 tons of carbon credits to offset the firm's recently calculated carbon footprint. Similarly, in The Carlyle Group's 2019 Corporate Sustainability Report, Co-Chief Executive Officers Youngkin and Lee celebrated their second consecutive year of operating on a net carbon-neutral basis, while reaffirming their commitment to ESG as an "essential way" of protecting capital from risk.[13]

Likewise, the Blackstone Group publicly declared their commitment to an ESG-centric approach. In his 2018 Annual Chairman's Letter, Co-Founder

Stephen Schwartzman made mention of his firm's ESG focus for the first time, while announcing the appointment of a new Global Head of ESG tasked with overseeing integration of ESG factors into investment decision-making across the firm.[14] That same year, Apollo Global Management published its first corporate ESG report—a decade after adopting its Responsible Investing Mission Statement.[15] Apollo tracks ESG data on its portfolio companies—totaling over 21,000 data points in 2018—and describes this practice as "essential" and as a "driver of value not solely through the lenses of risk management and compliance."[16]

Similarly, KKR Co-Founders Henry Kravis and George Roberts, in their 2018 Annual Letter to investors, stated their intention "to grow [their impact] investing business, which [they've] built based on the [ESG] efforts [they've] made over the last decade and on [their] history of investing in companies whose core business models seek to address global challenges."[17] 2018 also marked the launch of KKR's global impact fund, for which it raised over $1.1 billion by the close of 2019. KKR also led the private equity industry in 2009 by becoming the first of the top ten firms to join as a signatory of the UN Principles for Responsible Investment (UNPRI).[18] Three years later, Neuberger Berman followed suit.[19] The 2017 Neuberger Berman Annual Report states, "we have long believed that material [ESG] characteristics are an important driver of long-term investment returns, from both an opportunity and a risk-mitigation perspective."[20] The firm's 2019 report highlights that, of the firm's $339 billion of assets under management, 60% were managed with "consistent and demonstrable ESG integration."[21]

The evidence above suggests that ESG has evolved beyond its initial concept phase and now appears to be a priority agenda item across the private equity landscape—climbing out of obscurity into a period of comprehensive adoption. This rising tide of ESG interest is further demonstrated by the increase in ESG capital commitments and continued diversification of its investor base, beginning with wealthy individuals and family offices, and expanding to include foundations, endowments, pension funds, and other institutional investors. According to Preqin, private equity funds raised an estimated $595 billion in 2019, the third-highest annual total on record.[22] This fact, combined with the sudden rise in ESG capital commitments by UN Principles for Responsible Investment (UNPRI) signatories—many of whom are private equity managers—has shifted the industry's attention from "what is ESG?" to "how do we implement an ESG strategy?"

Characteristics of an ESG Strategy

PwC's 2019 Private Equity Responsible Investment Survey of 162 industry respondents showed that 91% have adopted or are currently adopting an ESG policy, 83% are concerned about climate risks to their portfolio, 81% report on ESG at the board level at least once per year, and 72% use or are developing metrics to track ESG performance.[23] PwC affirms that these statistics have been consistently rising since they first began surveying respondents on the subject in 2013.

The following represent the methods and tools that private equity firms often incorporate into their ESG strategy:

- *ESG Policy*—establishment of a statement or formal corporate policy that outlines a firm's approach and vision as it relates to ESG matters.
- *Diversity and Inclusion Policy*—adoption of a formal corporate policy that outlines a firm's approach and vision as it relates to diversity and inclusion of employees and other providers or stakeholders.
- *Industry Participation*—commitment to participate in ESG-related industry events including conferences and event sponsorships.
- *External Commitment*—supporting one or more ESG-principle organizations, such as the UN Principles for Responsible Investment (UNPRI) or the Global Impact Investment Network (GIIN) as a member, sponsor, advisor, or board member.[24]
- *Documentation*—written documentation of ESG-related matters, which may appear in internal documents, policy statements, investment memorandums, firm marketing materials, online data room, or website.
- *Distributed Publications*—distribution of ESG-related thought leadership and/or ESG-specific materials, such as company research, white papers, trade journals, online postings, newspapers, magazines, or other publications.
- *Staffing*—part-time or full-time hiring of internal staff or outside consultants dedicated to ESG-related matters on behalf of the firm. PwC's 2019 survey showed that 35% of private equity respondents report having a dedicated ESG team, up from 27% in 2016. Of those without dedicated ESG team members, 66% rely upon in-house deal teams to oversee ESG matters.
- *Process*—establishment of ESG-related procedures, often including due diligence, positive or negative screening, strategic divestment, thematic alignment, data gathering, research, reporting, investment decision-making, governance, or other firm activities that facilitate ESG efforts.

- *Product/Service*—establishment of an ESG investment strategy, sub-strategy, or other ESG-related financial products or services.
- *Transparency and Reporting*—incorporation of some form of ESG-related reporting, produced either in-house or via a specialized external reporting partner, which may or may not contain third-party sourced, verified, or audited data.
- *Mergers and Acquisitions (M&A)*—execution of one or more mergers or acquisitions of ESG-specialist firms in an effort to gain a competitive advantage or establish positioning as an ESG thought leader.[25]

While the above list is certainly not exhaustive, it is intended to highlight the most common characteristics of ESG strategies adopted by the private equity industry to date. As evidenced by the top ten private equity firms previously mentioned, firms choosing to incorporate ESG measures seek to meet demand for ESG, differentiate themselves from their peers, better align stakeholder interests, and most importantly, mitigate risk while maximizing returns, both financial and non-financial.

Importance of ESG Measurement in the Private Sector

Non-financial data measurement is central to an ESG strategy. Without it, financial data alone fails to give a comprehensive assessment of the true performance and embedded risks of an asset. For *environmental*, this may include energy or resource consumption data; for *social*, data pertaining to the job creation or supplier/procurement systems; and for *governance*, data related to employee diversity and remuneration. In 2018, the Institutional Limited Partners Association (ILPA) due diligence questionnaire—which is considered the industry standard in private equity—was revised to include a new section focused on ESG data gathering.[26] This further reinforces the belief that, without inclusion of ESG data in asset due diligence, non-financial performance is limited to a mere company byproduct rather than a strategic risk assessment tool or potential intentional source of profits.

For private equity managers, excluding this pertinent information in the decision-making process is illogical and can have negative implications on overall asset performance. McKinsey Quarterly states that, "companies that pay attention to [ESG] concerns do not experience a drag on value creation—in fact, quite the opposite."[27] It further finds that ESG is linked to value creation in five key ways: "(1) facilitating top-line growth, (2) reducing costs,

(3) minimizing regulatory and legal interventions, (4) increasing employee productivity, and (5) optimizing investment and capital expenditures."[28] For private equity managers to be successful in their ESG integration, it is of critical importance to decipher which ESG factors generate the greatest contribution to a firm's value.

Collecting this critical ESG data, however, remains a challenge in the private sector, particularly given the lack of reporting requirements by businesses and general partners. Data measurement hinges upon non-financial data availability, uniformity, accuracy, materiality, completeness, and reporting frequency—each of which is challenging in its own right. According to Morgan Stanley's 2018 Sustainable Signals Asset Owners Survey, 23% of investors identified the quality of ESG data as their top challenge.[29] Meanwhile, a 2019 Special Report issued by Private Equity International (PEI) demonstrated that 85% of limited partners considered ESG issues to be an important part of their investment decision-making process.[30]

Since industry-wide standards for ESG measurement and reporting still do not exist, several non-profit organizations have been established to provide supporting frameworks for disclosures, including:

- *UN Principles for Responsible Investment (UNPRI)*—Of the top 10 global private equity firms previously mentioned, four out of 10 are now UNPRI signatories, with KKR leading the group since 2009 and EQT, CVC Capital Partners, and Neuberger Berman following in later years.
- *UN Sustainable Development Goals (SDGs)*[31]—The 2019 PwC Private Equity Survey found that 67% of their 162 respondents stated that they have identified or are prioritizing the SDGs most relevant to their investments—up from 38% in 2016.
- *ESG Disclosure Framework for Private Equity.*[32]
- *Guidelines for Responsible Investment.*[33]
- *Global Impact Investing Rating System or GIIRS.*[34]

In addition to non-profit entities, a number of for-profit enterprises have emerged in recent years to offer ESG advisory and reporting services to the private equity industry. These independent firms often use client-specific methods or proprietary frameworks to produce ESG reports, as well as establish and support clients' internal ESG diligence processes.

This lack of standardization poses many challenges to the ESG assessments, monitoring, and reporting performed across the private equity industry today. However, countless studies repeatedly suggest there is an increasingly significant value placed on instituting an ESG-centric approach to investing by both limited partners and general partners.[35] Therefore, it seems reasonable for investment managers to continue to utilize whatever best-in-class ESG tools and methods are available to them, while standards and regulation catch up to demand.

ESG Challenges in Private Equity

Today, the use of ESG analysis in private equity is the racing equivalent of tire-rotation-while-fueling on the Formula 1[TM] track. Based on the growing global demand for sustainable investments, industry participants have rushed to digest and incorporate this two-pronged approach to investing. This mad dash to gain market share has caused unfortunate practices and revealed profound shortcomings that must be addressed in order for ESG's potential value to be respected.

The root cause of many of the challenges faced today stems from ESG's ambiguous and inconsistent definition. ESG is generally understood to be a set of tools or practices that illustrate or govern the non-financial performance of an investment. However, if you ask 10 private equity managers to define the scope of ESG factors, you will likely receive 10 unique responses.[36]

This definition confusion has led to a second major challenge of the private equity industry—determining the size of the ESG market, ranging from billions to trillions. While it may boil down to an explanation as simple—and as difficult to apply—as ESG being "in the eye of the beholder," it does a great disservice to the industry to suggest that the estimates for demand—or worse, actual deployment—far exceed their true figures. This not only presents a potential case of investment misrepresentation that legislation is lagging on, but it also poses a nightmare for ESG measurement and investment decision-making.

The Global Sustainable Investment Alliance is perhaps the most frequently-cited source for estimating a market size of $30 trillion.[37] It may be worth noting that this figure eclipses the total S&P 500 market cap of $21.42 trillion.[38] For further context, the global GDP, as last reported by the World Bank, stands at $85 trillion.[39] While it seems unlikely that there is more capital allocated to ESG than to the entire S&P index, its market definition and size remain fiercely debated, and consequently lack the clarity and validation the current private equity demand calls for.

For the time being, as private equity managers race to grab a seat at the sustainable investing table, it appears, in practice, acceptable to leave the scope and priority across ESG factors open to interpretation. Some investors have pursued relatively comprehensive and rigorous approaches to ESG implementation, while the majority have remained limited to more superficial and tactical messaging.

Analogous to miners flocking to a gold rush, it is not uncommon for firms to seek to convince the market of their long-standing history in the ESG field, which turns our attention to the challenge of revisionist history. In 2017, Neuberger Berman hired its first Head of ESG Investing, and yet their 80th anniversary annual report from 2019 states that, "Neuberger Berman started screening for ESG factors in its first decade." On the firm's webpage dedicated to ESG, it further states that 1989 marked their "[f]irst dedicated ESG-integrated strategy."[40] Even by the loosest definitions, both reported facts seem very unlikely. In 2018, Apollo Global Management published what they described as their "10th annual ESG report," though, curiously, all prior years are unavailable for public consumption.[41] This narrative of the sudden and miraculously extensive track record in ESG investing appears repeatedly throughout industry marketing materials and firm messaging. Numerous private equity firm websites are wallpapered with ESG rhetoric, suggesting that sustainability has long been of primary importance to the firm, when in reality, ESG funds generally make up single-digit percentages of overall assets under management, if at all. However, in their defense, this does not disprove that ESG factors may very well be active behind the scenes in shaping their policies, processes, and other investments, as is increasingly the case across the private equity landscape. ESG in private equity, while gaining momentum and maturity, is still lacking in much-needed regulatory support to maintain its integrity.

Case for Regulation in Promoting Sustainable Investments in Private Equity

Every nascent concept requires time and effort to ripen, though history has demonstrated that regulation and government support can accelerate this process by providing structure, standardization, and incentives, where they are otherwise lacking.

One area that seeks to benefit most from increased regulation is ESG metrics and reporting. Drawing a relevant comparison to the impact on public equities following the Stock Market Crash of 1929, the Securities Act of 1933 paved the way for the U.S. Securities and Exchange Commission (SEC) to adopt the Generally Accepted Accounting Principles as a financial reporting standard.[42] While there was significant backlash at the time, this reporting standardization provided uniform reporting expectations, while simultaneously protecting the interests of the shareholder—the latter being critical to the survival of the market. The hope is that ESG's critical mass today is sufficient incentive for regulation to present much-needed solutions before an event as apocalyptic as the Crash of 1929 forces government regulation within the private equity sector.

As it stands today, in absence of ESG regulation, the SEC has begun to investigate private equity funds whose ESG investment claims seem to diverge from their investment allocations and track records.[43] There is a strong need to separate signals from noise in the field of private equity sustainable investing. And yet, a crackdown preceding government ESG guidelines could protect the investor at the expense of seriously impeding ESG practices—which have, until this point, been an entirely voluntary undertaking.

Additionally, the size of a fund's position may make it extremely challenging to obtain ESG data from its portfolio companies, even if the metrics to be collected are clearly outlined. In this case, it would seem that regulation surrounding mandatory reporting requirements would ease the burden of the private equity manager. However, this also has the potential to shift the burden from the general partner to the portfolio company, which may not have the resources or ability to provide the required data. Therefore, it is imperative to consider the real impacts of regulation on all stakeholders.

Conclusion

> We decided long ago that the dangers of excessive and unwarranted concealment of pertinent facts far outweighed the dangers which are cited to justify it—John F. Kennedy, April 1961.

The rise of sustainable investing and the push for the best ESG metrics to support it demonstrate that the benefits of revealing pertinent facts far outweigh the benefits of concealing them. If history has taught us anything, it is that increased access to data leads to more informed decision-making—and that private investors are agile and strategically positioned to catalyze much-needed change.

Today, the facts point to a world that faces many urgent social and environmental challenges. Fortunately, the private equity industry has responded to this call for action through clear, demonstrable demand for solutions. By leveraging its strategic position, ESG tools have been established to support a rapidly evolving market for sustainable investments, which seek to efficiently deploy capital to investments that tie positive impact results to financial performance. The recent progress of the top 10 private equity firms incorporating ESG strategies serves as an example of the overall market's general direction. Demand for ESG and sustainable investment products is growing, with no sign of reversing course. While their individual approaches vary and challenges pertaining to execution remain, these are natural indications of a market reaching maturation. Following the classic adoption curve, it is expected that as more participants and capital enter the field, market standardization and consolidation will eventually take place.

For the time being, I offer that self-regulation, done well, may spare the private equity industry from government regulation that might be overly broad, value-destructive, and likely to retard necessary progress. In the interim, ESG measurement, reporting, and increasing implementation will, in time, yield a more efficient market. And while challenges are sure to remain, it is without a doubt that the private equity industry stands to benefit tremendously from the emergence and unparalleled growth of sustainable investments and the positive impact they aim to make.

Notes

1. Bain & Company. (2020). *Global Private Equity Rreport 2020*. Retrieved from https://www.bain.com/globalassets/noindex/2020/bain_report_private_e quity_report_2020.pdf.
2. Early history of private equity. *Wikipedia*. Retrieved from https://en.wikipedia. org/wiki/Early_history_of_private_equity.
3. Examples of such regulation may be found in the above-referenced endnote #2 and include the Small Business Investment Act of 1958, relaxing certain restrictions of the Employee Retirement Income Security Act (ERISA) of 1978, and putting forth the Economic Recovery Tax Act in 1981.
4. Tognini, Giacomo. (2020, April 1). Coronavirus Business Tracker: How the Private Sector Is Fighting the COVID-19 Pandemic. *Forbes*. Retrieved from https://www.forbes.com/sites/giacomotognini/2020/04/01/coronavirus-bus iness-tracker-how-the-private-sector-is-fighting-the-covid-19-pandemic/#530 d86fe5899.
5. International Finance Corporation. (2018). *Private Equity and Venture Capital's Role in Catalyzing Sustainable Investment*. Retrieved May 16, 2020 from https:// www.ifcamc.org/sites/amc/files/G20%20Input%20Paper_%202018.pdf.
6. International Finance Corporation. (2018). *Private Equity and Venture Capital's Role in Catalyzing Sustainable Investment*. Retrieved May 16, 2020 from https:// www.ifcamc.org/sites/amc/files/G20%20Input%20Paper_%202018.pdf.

7. From 2013 to 2018, $300 billion, $150 billion and $50 billion of private equity capital was invested in sustainable investments in the United States, Western Europe, and emerging markets, respectively—further demonstrating private equity's catalytic role in sustainable investing.

8. Friede, G., Busch, T., & Bassen, A. (2015). ESG and Financial Performance: Aggregated Evidence from More Than 2000 Empirical Studies. *Journal of Sustainable Finance & Investment, 5*(4), 210–233. https://doi.org/10.1080/204 30795.2015.1118917.

9. PRI. (2020). What Are the Principles for Responsible Investment? *PRI.* Retrieved from https://www.unpri.org/pri/an-introduction-to-responsible-inv estment/what-are-the-principles-for-responsible-investment.

10. Friede, G., Busch, T., & Bassen, A. (2015). ESG and Financial Performance: Aggregated Evidence from More Than 2000 Empirical Studies. *Journal of Sustainable Finance & Investment, 5*(4), 210–233. https://doi.org/10.1080/204 30795.2015.1118917.

11. Kolakowski, M. (2020, March 2). World's Top 10 Private Equity Firms. *Investopedia.* Retrieved from https://www.investopedia.com/articles/markets/011116/ worlds-top-10-private-equity-firms-apo-bx.asp.

12. Bloomberg Surveillance. (2020, January 20). Billionaire Robert Smith on Workforce Challenges ESG. *Bloomberg.* Retrieved from https://www.bloomb erg.com/news/videos/2020-01-20/billionaire-robert-smith-on-workforce-challe nges-esg-investing-video.

13. The Carlyle Group. (2019). *Advancing the Art of What's Possible: Corporate sustainability Report,* pp 2–3. Retrieved from https://www.carlyle.com/sites/def ault/files/reports/carlyleccr2019.pdf.

14. Schwartzman, S. A. (2018). *Performance and Innovation: Blackstone's Chairman's Letter 2018.* Retrieved from https://ir.blackstone.com/sec-filings-annual-letters/ default.aspx.

15. Apollo Global Management. (2018). *The ESG Opportunity: How Apollo Drives Change—2018 ESG Summary Annual Report,* pp 1–3. Retrieved from https:// www.apollo.com/~/media/Files/A/Apollo-V2/documents/apollo-2018-esg-sum mary-annual-report.pdf.

16. Apollo Global Management. (2018). *The ESG Opportunity: How Apollo Drives Change—2018 ESG Summary Annual Report.* Retrieved from https://www.apo llo.com/~/media/Files/A/Apollo-V2/documents/apollo-2018-esg-summary-ann ual-report.pdf.

17. KKR and Company. (2018). *KKR 2018 Annual Letter: Connecting the Dots,* p. 3. Retrieved from https://ir.kkr.com/static-files/dcc93755-6284-4b0f-9e87- 7f786e38924a.

18. PRI. (2020). PRI Signatory Directory: Kohlberg Kravis Roberts & Co. L.P. *PRI.* Retrieved from https://www.unpri.org/signatory-directory/kohlberg- kravis-roberts-and-co-lp/1391.article.

19. PRI. (2020). PRI Signatory Directory: Neuberger Berman Group LLC. Retrieved from https://www.unpri.org/signatory-directory/neuberger-berman-group-llc/1533.article.

20. Neuberger Berman. (2017). *Neuberger Berman 2017 Annual Report*, p. 4. Retrieved from https://www.nb.com/-/media/NB/Firm-Pages/Annual-Reports/2017/Neuberger_Berman_Annual_Report_2017.ashx.

21. Neuberger Berman. (2019). *Celebrating Neuberger Berman at 80 Years*, pp. 6. Retrieved from https://www.nb.com/handlers/documents.ashx?id=fd3 e161b-d49a-434b-a9b3-65f9cf39e2b0&name=Celebrating_NB_80_Years.

22. Preqin. (2020, January 9). *2019 Private Equity & Venture Capital Fundraising & Deals Update*. https://www.preqin.com/insights/special-reports-and-factsheets/2019-private-equity-venture-capital-fundraising-deals-update/26639.

23. Jackson-Moore, W., Case, P., Bobin, E., & Janssen, J. (2019). *Older and Wiser: Is Responsible Investment Coming of Age? Private Equity Responsible Investment Survey 2019*, pp. 1–9. Retrieved from https://www.pwc.com/gx/en/services/sus tainability/assets/pwc-private-equity-responsible-investment-survey-2019.pdf.

24. Global Impact Investing Network. (2020). About the GIIN. *Global Impact Investing Network*. Retrieved from https://thegiin.org/about/.

25. It should be noted that mergers and acquisitions are generally found in more mature markets and, therefore, are not (yet) prevalent among private equity institutions. This may change as the market continues to develop.

26. Institutional Limited Partners Association. (2018). *Due Diligence Question-naire, Version 1.2*. Retrieved from https://ilpa.org/wp-content/uploads/2018/09/ILPA_Due_Diligence_Questionnaire_v1.2.pdf.

27. Koller, T., Nuttal, R., & Henisz, W. (2019). Five Ways That ESG Creates Value. *McKinsey Quarterly*. Retrieved from https://www.mckinsey.com/bus iness-functions/strategy-and-corporate-finance/our-insights/five-ways-that-esg-creates-value.

28. Koller, T., Nuttal, R., & Henisz, W. (2019). Five Ways That ESG Creates Value. *McKinsey Quarterly*.

29. Morgan Stanley & Co. LLC. (2018). *Sustainable Signals: Asset Owners Embrace Sustainability*. Retrieved from https://www.morganstanley.com/assets/pdfs/sus tainable-signals-asset-owners-2018-survey.pdf.

30. Private Equity International. (2018, December). *PEI Perspectives 2019 Special Report*. Retrieved June 18, 2020 from https://www.morganstanley.com/assets/pdfs/sustainable-signals-asset-owners-2018-survey.pdf.

31. United Nations Development Programme. (2020). Sustainable Development Goals. *United Nations Development Programme*. Retrieved from https://www.undp.org/content/undp/en/home/sustainable-development-goals.html.

32. The Association for Private Capital Investment in Latin America. (2013). *Envi-ronmental, Social, and Corporate Governance (ESG) Disclosure Framework for Private Equity*. Retrieved from https://lavca.org/wp-content/uploads/2017/02/13161_ESG_Disclosure_Document_v6.pdf.

33. American Investment Council. (2009). Guidelines for Responsible Investing. *American Investment Council.* Retrieved from https://www.investmentcouncil. org/guidelines-for-responsible-investing/.

34. B Lab. (2020). Company Ratings. *B Analytics.* Retrieved from https://b-analyt ics.net/content/company-ratings.

35. Indahl, R., & Jacobsen, H. (2019). Private Equity 4.0: Using ESG to Create More Value with Less Risk. *Journal of Applied Corporate Finance, 31*(2), 34–41. https://doi.org/10.1111/jacf.12344.

36. Esty, D. & Cort, T. (2017). State of ESG Data and Metrics. *Journal of Environmental Investing, 8*(1). Retrieved from http://www.thejei.com/wp-content/ uploads/2017/11/Journal-of-Environmental-Investing-8-No.-1.rev_-1.pdf.

37. Global Sustainable Investment Alliance. (2019). *2018 Global Sustainable Investment Review.* Retrieved from http://www.gsi-alliance.org/wp-content/uploads/ 2019/03/GSIR_Review2018.3.28.pdf.

38. Ycharts. (2020, April 20). S&P 500 Market Cap as of March 2020. *Ycharts.* Retrieved from https://ycharts.com/indicators/sp_500_market_cap

39. World Bank. (2020). GDP (Current US$). *World Bank DataBank.* Retrieved from https://data.worldbank.org/indicator/NY.GDP.MKTP.CD.

40. Neuberger Berman. (2020). Environmental, Social and Governance Investing: ESG Philosophy. *Neuberger Berman.* Retrieved February from https://www.nb. com/en/global/esg/philosophy.

41. Apollo Global Management. (2018). *The ESG Opportunity: How Apollo Drives Change—2018 ESG Summary Annual Report.*

42. Neuberger Berman. (2019). *Celebrating Neuberger Berman at 80 Years.*

43. Chung, J., & Michaels, D. (2019, December 16). ESG Funds Draw SEC Scrutiny. *The Wall Street Journal.* Retrieved from https://www.wsj.com/articles/ esg-funds-draw-sec-scrutiny-11576492201.

10

Avoiding the Tragedy of the Horizon: Portfolio Design for Climate Change-Related Risk Management and the Low-Carbon Energy Transition

Jennifer Bender, Todd Arthur Bridges, Kushal Shah, and Alison Weiner

Abstract This chapter discusses recent developments in how institutional investors are approaching the integration of climate change considerations into their equity portfolios. In recent years, increasing evidence of climate change impacts combined with emerging GHG regulatory requirements has generated rapid innovation in investing with an eye toward carbon exposure. No longer confined to small satellite allocations in their portfolios, investors are seeking integration methods at the total portfolio. We discuss a range of portfolio design approaches in this chapter from simple screening-based approaches, which have been prevalent within ESG investing for some years, to newer optimization-based portfolio construction techniques embedding multiple climate change metrics and criteria.

Keywords Portfolio design · ESG disclosure requirements · Carbon exposure · Transition risk · Physical climate-related risk · Green investments · Climate change resilient investments · Brown revenue · Green

J. Bender (✉) · K. Shah
State Street Global Advisors, Boston, MA, USA
e-mail: Jennifer_Bender@ssga.com

T. A. Bridges
Arabesque, Boston, MA, USA

A. Weiner
Capital Group, New York, NY, USA

© The Author(s) 2020
D. C. Esty and T. Cort (eds.), *Values at Work*,
https://doi.org/10.1007/978-3-030-55613-6_10

143

revenue · Carbon data · Fossil fuel reserves · Sustainability screening · Mitigation approach · Climate-aware portfolio

Major developments have helped shift the climate change conversation from debate to action, creating a new sense of urgency among investors. These developments include more frequent manifestations of climate change impacts, such as record-breaking temperatures in many parts of the world and increased frequency of natural disasters, e.g., wildfires, hurricanes, and flooding. Governments are devoting increasing resources to analyze the current and potential economic and financial losses associated with these manifestations. Regulatory bodies are responding—building on the Paris Climate Change Agreement. And legislation is evolving, particularly in Europe. Last but not least, voluntary initiatives are growing in support of broader corporate action on climate change, such as the Task Force on Climate-related Financial Disclosures (TCFD).

Institutional investors are playing a part in this quickly evolving world of climate-change response, driven by concerns over the potential impact of climate change to their investments, regulatory pressure, and in some cases, a growing desire for organizations to align their portfolios with broader priorities. This chapter provides an overview of our answers to the four core questions of climate-aware investing:

1. What are the risks and opportunities posed by climate change in my portfolio?
2. What data is available to quantify these risks and opportunities?
3. What portfolio design/construction approaches are available for investors integrating climate change considerations?
4. What are the performance impact implications of building a climate-aware portfolio?

Mark Carney, former Governor of the Bank of England and current Chairman of the Financial Stability Board, has said:

> *Climate Change is the Tragedy of the Horizon.* We don't need an army of actuaries to tell us that the catastrophic impacts of climate change will be felt beyond the traditional horizons of most actors—imposing a cost on future generations that the current generation has no direct incentive to fix.

Long-term investments are made against this nearing horizon, and investors focused on long-term returns and risks have an imperative to give due consideration to climate change-related threats and opportunities.

Recent Developments in Investment, Climate Change-Related Regulations, and Voluntary Initiative

Before we tackle the four key questions above, it is worth providing an overview of recent developments in the global investment community. First, government action, particularly in Europe, is taking place to address the commitments made through the Paris Agreement. The European Commission has focused on how to achieve the promised climate and energy targets of a 40% reduction of greenhouse gas emissions, 27% renewables in energy consumption, and 30% energy savings by 2030. In 2018, the European Commission published an *action plan* which outlines legislative initiatives related to a number of items: establishment of a European Union classification system or taxonomy, disclosure requirements for a range of market participants, and new measures regarding benchmarks.

Regarding benchmarks, in 2019, the European Unions's Technical Expert Group (TEG) released a comprehensive set of recommendations for EU Climate Transition Benchmarks (EU CTBs) and EU Paris-Aligned Benchmarks (EU PABs). Criteria for meeting these benchmarks include a set level of carbon reduction (30% for Climate Transition Benchmarks and 50% for Paris-Aligned Benchmarks) and exclusions on fossil fuels and electricity producers with high greenhouse gas emissions. Final rules are expected sometime in 2020.[1]

In addition, individual countries are also advancing legislation. For instance, in France, asset managers with more than 500 million euros under management must disclose how they assess climate change-related portfolio risks. In the United Kingdom, government authorities announced in 2019 their intention to explore a mandatory requirement for listed companies and pension funds to disclose climate change-related risks which might take place as early as 2022. The Bank of England subsequently unveiled at the end of 2019 plans to introduce a mandatory and uniform climate risk test for major banks and insurers in 2021.

At the same time as government action, investors are coalescing around voluntary frameworks such as the UN Principles for Responsible Investment (UN PRI) and the Financial Stability Board's Task Force on Climate-related Financial Disclosures. Signatories to the UN Principles for Responsible Investment represent a global investor base with more than 2450 investors and over $82 trillion in assets under management.[2] They have collectively pledged to incorporate environmental, social, and corporate governance issues into investment practices and promote disclosure by companies they invest in. Meanwhile, the Task Force on Climate-related Financial Disclosures, a market-driven initiative established in 2016, has made rapid progress by enlisting over 700 supporters including large asset owners and managers, as well as select government entities.

Investment Risks and Opportunities Posed by Climate Change

The first question we now consider is, what are the investment risks and opportunities posed by climate change? Climate change poses multiple risks and opportunities to investment portfolios. The risks include physical and transition risk, while opportunities include investments in green energy and in companies that are building climate change resiliency into their businesses.

Physical and Transition Risks

Physical risks are tangible risks of climate change that could manifest through a rise in sea levels, droughts, flooding, extreme temperatures, and increased frequency of extreme weather events. These phenomena could damage infrastructure, cause supply-chain disruption, result in raw materials scarcity, or harm human health. These climate change-related risks have the potential to transmit across economic and financial systems, with some—such as banks, (re)insurance companies, agriculture, and consumer goods—experiencing particularly significant shocks.

Transition risks are risks to economic and business models that are associated with changing government policy and/or changing consumer and investor preferences. On the one hand, transition risks may arise from changing government policies and regulations including, for example, new carbon pricing, fossil fuel taxes, or emissions trading schemes leading to higher costs for carbon emissions and the likelihood of stranded assets. Transition risks also include changing consumer habits and labor market shifts, as well as investment allocation decisions in companies and sectors better suited to a low-carbon economy. For example, investors may reposition investments away from fossil fuel-reliant sectors toward "green" revenue sectors, particularly in energy generation, energy equipment, energy management, energy efficiency, environmental infrastructure, and environmental resources.

Opportunities

What are the investment opportunities associated with climate change? Green investments represent an opportunity to reposition toward companies growing their share of "green" revenue—the percentage of sales of products that benefit the environment, such as with cleaner air, water, or land. These companies include renewable energy companies such as those involved in wind-generated power, solar power, etc. They also include companies that are focused on improving energy management and efficiency and environmental infrastructure. Green revenues can also exist in traditional manufacturing industries—e.g., eco-friendly consumer products and electric vehicles.

Another potential opportunity is climate change-resilient investments, firms which are poised to be successful because they are building in resiliency to the potential impacts of climate change. These companies are changing their business models, including how they deploy capital and where they grow their business lines, in response to climate change. These are firms that are in some sense "adapting" to climate change, particularly the potential physical impacts of climate change, by insulating themselves against risks such as rising coastlines, extreme weather events, changing precipitation, and water supplies. These physical risks essentially become potential new opportunities for investors.

Data Is at the Center of Climate Change-Aware Investing

Given the risks and opportunities investors are being confronted with, the next question is: what data is available to quantify these risks and opportunities?

The Data Landscape Is Growing Quickly

An explosion of climate change data in recent years has made it possible for robust investable solutions to be built around these themes. Supported by the disclosure efforts of the Carbon Disclosure Project (CDP), carbon data is now available from multiple commercial providers, including S&P, Morgan Stanley Capital International (MSCI), Sustainalytics, Institutional Shareholder Services, and Bloomberg, to name a few. Carbon data comes in many forms including metrics capturing different types of greenhouse gas emissions (e.g., Scope 1, Scope 2, Scope 3) as well as different units (e.g., Weighted Average Carbon Intensity, Total Carbon Emissions, Carbon Footprint, etc.).

Carbon data and fossil fuel reserves are typically sourced by data vendors directly from companies in annual reports, Corporate Social Responsibility (CSR) reports, environmental reports, and company websites. In addition to sourcing the formally disclosed data, data providers have research teams that consolidate the raw data and enhance the data quality and coverage through company validation checks, data standardization, and reporting sequencing (time lags). Given that company carbon disclosure is still voluntary, disclosure rates vary by global region and it is difficult to systematically capture the environmental impacts in quantitative terms. As a result, carbon data providers have developed unique, and frequently proprietary, estimation methodologies—based on industry-specific regression models—to allow for increased data coverage that is comparable between companies. In addition to carbon data, other forms of data capturing climate change-related risks include fossil fuel reserves (e.g., the amount of fossil fuel reserves owned by a company), fossil fuel sector exposure (e.g., a company's exposure to the fossil fuel-based economy), and *brown* revenue (e.g., the portion of a company's revenue derived from high-carbon/fossil fuel sources).

A corollary to brown revenue, on the opportunity side, is *green* revenue, which is the portion of a company's revenue derived from green goods, products, and services. Green revenue is generally more difficult to quantify and requires company-level assessments by trained analysts. For instance, the Financial Times Stock Exchange employs a proprietary industrial taxonomy covering 64 subsectors to categorize company revenues as green or non-green.

Lastly, climate change resilience is among the most difficult dimensions to quantify, which data providers are only beginning to address. For example, Institutional Shareholder Services produces a rating on whether a company has a clear position on climate change (i.e., position on the scientific evidence of climate change, the company's responsibility in this context and its commitment to contribute to the reduction of greenhouse gas emissions). Other companies are starting to offer metrics that assess the climate orientation of capital expenditures and company business lines. These nascent datasets offer a compelling source of information still to come (Table 10.1).

Data Challenges

Despite the rapidly growing number of data sets capturing different dimensions of climate change risk and opportunity, there are significant challenges related to the data. First and foremost is that the data is largely estimated

Table 10.1 Sample climate/environmental data

	Metric	Description
Carbon data	Carbon-Scope 1 (tons carbon dioxide equivalent)	Greenhouse gas emissions generated from burning fossil fuels and production processes which are owned or controlled by the company[a]
	Carbon-Scope 2 (tons carbon dioxide equivalent)	Greenhouse gas emissions from consumption of purchased electricity, heat, or steam by the company[b]
	Carbon Intensity-Direct + First Tier Indirect (tons carbon dioxide equivalent/$ million)	Greenhouse gases emitted by the direct operations of and suppliers to a company divided by revenue
Fossil fuel reserves	Total Reserves carbon dioxide emissions (million tonnes)	Total embedded greenhouse gas emissions for the company in financial year
Green revenue	Revenues from green or low-carbon technology products/business	Percentage of revenues from green or low-carbon technology products/business
Brown revenue	Revenues from brown sectors	Percentage of revenues from activities related to extraction of fossil fuels, or power generation from fossil fuels

[a]World Resources Institute & World Business Council for Sustainable Development. (2020). *Greenhouse Gas Protocol.* Retrieved from https://ghgprotocol.org/
[b]World Resources Institute & World Business Council for Sustainable Development. (2020). *Greenhouse Gas Protocol*

and not self-reported. That is, mandatory consistent reporting, along the lines of traditional balance sheets and income statements, does not currently exist anywhere in the world. Even for carbon emission data, arguably the furthest along in terms of disclosure rates, only about 70% of companies in the Morgan Stanley Capital International World Index report actual carbon data.[3] Moreover, even the companies who do report may be under-reporting. Studies suggest significant gaps in emissions disclosure among European, Middle Eastern, and African companies[4] and North American companies. For instance in North America, of 1900 companies that disclosed their metrics to the Carbon Disclosure Project, the companies under-report their carbon emissions by 7% on average.[5]

In addition to the estimation and disclosure issues, many data modeling questions remain. Focusing on carbon emissions data, some examples include:

- **Scope**: Carbon emissions are generally classified into three categories—Scope 1, 2, and 3.[6] Scope 3 is viewed by many as the most important of the three because it is the most expansive; however, it is also the most difficult to measure. The Task Force on Climate-related Financial Disclosures's 2019 disclosure recommendations focused on Scope 1 and 2 only, because of these measurement challenges.
- **Normalization of Emissions at the Company Level**: Currently, there are various ways to normalize carbon—by sales, by market cap, by revenue, etc. No one method has gained broad-based acceptance by the industry to date. The Task Force on Climate-related Financial Disclosures's 2019 recommendations do include a measure scaled by revenue (e.g., Weighted Average Carbon Intensity) as well as multiple measures scaled by market capitalization (e.g., Carbon Footprint and Total Emissions).
- **Missing Data**: While coverage for carbon data is generally high from commercial providers, there are still segments of securities with missing data, particularly securities of smaller market capitalization and those in emerging markets. Typical data treatments such as filling missing values with sector averages, industry averages, or other means, can have a significant impact on the treated values.
- **Historical Data Availability**: Security-level historical data is constrained by the relative newness of environmental, social, and governance (ESG) data and carbon data. For instance, reasonable coverage (i.e., 90%+ securities for standard universes such as the S&P 500, Russell 1000, and Morgan Stanley Capital International World Indices) for carbon emissions generally starts in 2009.
- **Retroactive Revisions to Data**: Because carbon data is largely estimated, restatements of carbon data occur much more frequently than for typical financial data.

These sample data issues are illustrative of the broader issues around climate/carbon data, a challenge that is emblematic of all ESG-related data. Until disclosures improve along a standard reporting framework, these challenges will remain for any investor seeking to integrate climate change considerations into their portfolio.

Integrating Climate Change Considerations into Investment Portfolios

The third question we tackle is around the portfolio design and construction approaches available for investors integrating climate change considerations. Climate change-*aware* investing has been growing for almost a decade but only recently has research around the intersection of climate and traditional portfolio management taken off. Traditional investment theory largely centers around portfolio return and risk, and the linkage between climate metrics, return, and risk is constrained by the relative newness of climate data. That said, there are at least three approaches that have gained broader adoption:

Screens-Based Approach

A screens-based approach has historically been the most common way investors have incorporated ESG considerations into their portfolios. Within the realm of climate investing, screening entails removing exposure to specific high-carbon-emitting industries such as utilities or traditional fossil fuel-related industries, such as coal and petroleum producers. Investors may pursue this approach with the goal of avoiding stranded-asset risk, or may view investments in specific industries as a drag on portfolio performance in the long run.

Screened strategies can be viewed as addressing both the physical and transition risk of climate change because they often involve removing exposure to the fossil fuel industry, which may be disproportionately impacted by both those risk categories. These strategies are particularly favored by institutions with actively engaged beneficiaries or other stakeholders who seek to minimize involvement with major contributors to climate change.

Mitigation Approach

Mitigation approaches set an explicit objective to reduce the flow of heat-trapping greenhouse gases into the atmosphere and increase exposure to "green" companies. Investors primarily seeking to manage the economic impacts of climate change across their portfolio more commonly pursue mitigation.

Mitigation strategies can be viewed as addressing the same risks as screened strategies in a more targeted way (often with a targeted level of carbon reduction), with the option of also adding exposure to opportunities such as green revenue. Specifically, these portfolios allow investors to manage the transition risk of being exposed to high emitters during the shift to a low-carbon economy, including changes in the value of carbon-intensive assets or carbon-emitting activities due to regulation. This mitigation objective is accomplished by reducing carbon intensity and brown revenue exposure in a portfolio while increasing exposure to "green" businesses.

Because the mitigation approach may have multiple objectives and data inputs, more sophisticated portfolio construction approaches are often needed. Depending on how many metrics there are and whether an explicit target is required (e.g., a carbon reduction target of 50%), rules-based approaches such as simple screening and tilted approaches may not be precise enough, and algorithms such as optimization-based approaches may be necessary.

Mitigation and Adaptation Approach

Mitigation and adaptation strategies allow investors to respond comprehensively across climate risks and opportunities by combining the benefits of a mitigation approach with exposure to both green revenue and climate change-resilient businesses. This approach extends mitigation to include an explicit objective to increase exposure to companies adjusting to actual or expected future climate change impacts. Similar to the mitigation approach, there are multiple objectives and data inputs in mitigation and adaptation strategies, and more sophisticated portfolio construction approaches involving quantitative algorithms are usually needed.

These three portfolio design and construction approaches are not mutually exclusive. It is possible, for example, to apply both screening and a mitigation approach to a portfolio.

Performance Implications of Climate-Aware Portfolios

Finally, what are the performance impacts of building a climate-aware portfolio? As we previously noted, the available history of climate metrics is limited. That said, we do have at least five years of reasonable data coverage for most metrics—and 10 years of reasonable data coverage for carbon-related measures—and can achieve some sense of the likely impact on certain representative portfolios.

We illustrate the performance implications by considering a representative set of broad global equity portfolios, closely mirroring a typical institutional investor's policy benchmark (e.g., the Morgan Stanley Capital International World Index). In this set of experiments, we simulate the historical returns and risks of portfolios designed along the screening, mitigation, and mitigation and adaptation approaches described in the previous section.

For illustrative purposes, we focus on building portfolios that are aligned with the recent recommendations from the EU Technical Expert Group around the target level of carbon reduction necessary to meet the standards for EU Climate Transition Benchmarks and EU Paris-Aligned Benchmarks, described earlier in the chapter. The portfolios we create are as follows:

Screened Portfolios

- Screened Portfolio #1 (remove worst 2.5% polluters): Approximate 30% *carbon reduction* (EU Climate Transition Benchmark-aligned) achieved through screening out the worst 2.5% polluters, measured by *carbon intensity*
- Screened Portfolio #2 (remove worst 10% polluters): Approximate 50% *carbon reduction* (EU Paris-Aligned Benchmark-aligned) achieved through screening out the worst 10% polluters, measured by *carbon intensity*
- Screened Portfolio #3 (remove carbon-intense sectors) Approximate 50% *carbon reduction* (EU Paris-Aligned Benchmark-aligned) through the removal of the Energy, Materials, and Utilities sectors, which are the three highest carbon-emitting sectors.

Mitigation

- Mitigation Portfolio #4 (weighting full portfolio to reduce 30% carbon emissions): Explicit 30% Carbon Reduction (EU Climate Transition Benchmark-aligned) achieved by using optimization to achieve the specified level of carbon intensity relative to the market cap-weighted benchmark
- Mitigation Portfolio #5 (weighting full portfolio to reduce 50% carbon emissions): 50% Carbon Reduction (EU Paris-Aligned Benchmark-aligned) achieved by using optimization to achieve the specified level of *carbon intensity* relative to the market cap-weighted benchmark.

Mitigation and Adaptation

- Mitigation and Adaptation Portfolio #6 (weighting full portfolio with additional reduction of fossil fuels to reduce 50% carbon emissions): 50% *carbon reduction* (EU Paris-Aligned Benchmark-aligned) with additional climate change objectives around reducing fossil fuels and brown revenue exposure, and improving green revenue exposure, and adaptation ratings achieved by using optimization to balance these multiple objectives.

To illuminate the characteristics of portfolios which take these different approaches, we modeled the six portfolio cases described above and compared their attributes and performance against the Morgan Stanley Capital International (MSCI) World Index benchmark. The performance summary of the six hypothetical cases (Exhibit 2) demonstrates the possible alignment of climate change concerns and investment objectives.

Key observations from our simulations are as follows:

- There is a possibility of achieving a low tracking error[7] in climate change-aware portfolios. For example, using a mitigation strategy to reduce carbon intensity by 50% (Mitigation Portfolio #5) we saw only 19 basis points of tracking error. When also reducing fossil fuel exposure (Mitigation and Adaptation Portfolio #6), the tracking error increases to only 32 basis points while also achieving substantive improvements in green revenue exposure and climate risk adaptation alongside reductions in carbon intensity, fossil fuels reserves exposure, and brown revenue exposure.
- Active returns for all six cases are small but positive, in line with low-targeted levels of tracking error. Sharpe Ratios[8] are equivalent to the benchmark suggesting a similar risk-weighted return. While returns are usually not the primary focus of climate-oriented investors, the results imply that climate change goals can be achieved without substantial loss of financial performance and with potentially even better performance.
- Screening can be used to meaningfully reduce carbon intensity but generally results in higher levels of tracking error for the same amount of carbon intensity, compared to optimization-based approaches. To achieve 50% carbon intensity reduction using a screening method (Screened Portfolio #2) results in 43 basis points of tracking error, compared to the Mitigation Portfolio #5, which results in the same carbon intensity reduction (50%), but with only 19 basis points of tracking error. Moreover, screening out industries, as opposed to individual stocks, using carbon intensity can

lead to higher levels of tracking error without a commensurate reduction in carbon intensity. This result is exhibited by Screened Portfolio #3, where the exclusion of companies in the energy, materials, and utilities sectors results in a tracking error of 138 basis points with only a modest incremental reduction in carbon intensity. Thus, the amounts of carbon reduction achieved through screening out the worst polluting companies (Screened Portfolios #1 and #2) can be achieved at a lower tracking error using mitigation approaches where all companies are included in the portfolio, but are weighted differently based on carbon intensity (Mitigation Portfolios #4 and #5).

- Screening-based approaches result in a loss of green revenues. This is due to the exclusion of materials and utilities sectors, which have a higher concentration of green companies.

Overall, the results show that the way climate change-related goals are integrated into investment portfolios should reflect the chosen investing criteria and desired performance measures. Moreover, as previously noted by the authors, any specific set of climate change goals can be achieved in a variety of different ways (Table 10.2).[9]

Summary

Investors are increasingly aware of the potential systemic risks climate change poses to the financial system. As a result, they are increasingly integrating climate change-related data into their portfolios. Research into climate-aware investment solutions is expanding rapidly as new data sets become available that quantify both the potential risks from climate change and the opportunities that are being created.

Already, we are witnessing a number of ways investors are addressing climate change in their portfolios. Screening-based or exclusionary approaches have been the most widely used approach for building climate-aware investment portfolios. Fossil fuel divestment, for instance, gained popularity in the 2000s, generating high-profile headlines, before experiencing a backlash in the 2010s because of its impact on financial performance. Screening is still widely used today, but it has migrated toward targeting certain companies such as thermal coal producers, power generators, companies involved in Arctic drilling, and other companies directly involved in carbon emissions growth. Mitigation and mitigation/adaptation are newer approaches that allow investors to achieve multiple objectives,

Table 10.2 Detailed results of simulated climate change-aware portfolios[a]

	Morgan Stanley Capital International World Index (World Index)	Screened Portfolio #1 (Screened approx. 30% carbon reduction)	Screened Portfolio #2 (Screened approx. 50% carbon reduction)	Screened Portfolio #3 Energy, materials, industrials excluded	Mitigation Portfolio #4 Mitigation (30% carbon reduction)	Mitigation Portfolio #5 Mitigation (50% carbon reduction)	Mitigation and Adaptation Portfolio #6 Mitigation and adaptation
Annual return (%)	9.94	9.99	10.25	11.09	10.05	10.07	10.3
Annual vol (%)	11.46	11.55	11.51	11.54	11.44	11.42	11.43
Sharpe ratio	0.87	0.86	0.89	0.96	0.88	0.88	0.9
Active return (%)	–	0.05	0.31	1.15	0.11	0.13	0.36
Tracking error (%)	–	0.19	0.43	1.38	0.14	0.19	0.32
Information ratio	–	0.26	0.72	0.84	0.78	0.7	1.15
Maximum drawdown (%)	–13.31	–13.52	–13.43	–13.16	–13.28	–13.16	–13.13
Beta	–	1.01	1	1	1	1	1
Number of stocks (avg)	1636	1595	1473	1326	1078	1030	928

	Morgan Stanley Capital International World Index	Screened Portfolio #1 Screened (approx. 30% carbon reduction)	Screened Portfolio #2 Screened (approx. 50% carbon reduction)	Screened Portfolio #3 Energy, materials, industrials excluded	Mitigation Portfolio #4 Mitigation (30% carbon reduction)	Mitigation Portfolio #5 Mitigation (50% carbon reduction)	Mitigation and Adaptation Portfolio #6 Mitigation and adaptation
Effective no. of stocks (avg)	350	341	313	282	347	336	325
Carbon intensity (% improvement)	–	32.2	55.3	65	30	50	50
Fossil fuel reserves (% improvement)	–	–1.2	26.8	99.2	–3.7	–4.2	50
Brown revenues (% improvement)	–	9.5	36.9	97.5	11.2	21.1	50

[a]Results in Exhibit 2 are based on data from July 2013 to July 2019 using $ Gross Returns. The companies in the analysis were drawn from the MSCI World Index constituents. The benchmark portfolio is the MSCI World Index. Performance data was calculated using a monthly rebalancing frequency

including, most importantly, traditional risk objectives such as minimizing tracking error and concentration.

As climate considerations and sustainable investing issues become increasingly important for all types of investors, we expect to see further innovations in data integration, portfolio-level techniques, and climate finance research. While government-led regulations will play a role, institutional investors who are focused on long-term returns and risks have a growing incentive to care about the impact of climate change because of its importance to investable assets. This group of investors is likely to play a leading role in avoiding the tragedy of the horizon.

Notes

1. EU Technical Expert Group. (2019). *Report on Benchmarks*. European Union.
2. UN PRI. (2019). *The Inevitable Policy Response*. United Nations.
3. Calculated using Scope 1 disclosure information from S&P Trucost, as of 30 September, 2019.
4. S&P Dow Jones Trucost. (2018a). *Corporate Carbon Disclosure in Europe, the Middle East, and Africa* (White paper).
5. S&P Dow Jones Trucost. (2018a). *Corporate Carbon Disclosure in North America* (White paper).
6. Scope 1 emissions are direct emissions from sources that are owned or controlled by the company and include for instance on-site fossil fuel combustion and fleet fuel consumption. Scope 2 emissions are indirect emissions from sources that are owned or controlled by the company and include emissions that result from the generation of electricity, heat, or steam purchased from a utility provider. Scope 3 emissions are from sources not owned or directly controlled but related to the company's activities. They include emissions generated by a company's non-electricity supply chain, employee travel and commuting, and emissions associated with contracted solid waste disposal and wastewater treatment. Scope 3 is often referred to as "upstream" emissions.
7. Tracking error is the divergence between the price behavior of a position or a portfolio and the price behavior of a benchmark. It is a mechanism to compare a potential portfolio against a target for financial performance.
8. A Sharpe Ratio measures the average return of the portfolio minus the risk-free return divided by the standard deviation of return on an investment. It is a measure of risk-weighted return.
9. Bender, J., Bridges, T. A., & Shah, K. (2019). Reinventing Climate Investing: Building Equity Portfolios for Climate Risk Mitigation and Adaptation. *Journal of Sustainable Finance & Investment, 9*(3), 191–213.

Part IV

Regulations, Liability, and Fiduciary Duty of ESG Disclosure

11

Recent Developments in ESG Reporting

Paul A. Davies, Paul M. Dudek, and Kristina S. Wyatt

Abstract Environmental, social, and governance (ESG) issues have become a core business concern and a central focus of investors. The ESG reporting landscape is fragmented, with hundreds of reporting standards forming what has been called an alphabet soup of acronyms with no common framework to guide corporate disclosures. Companies' ESG disclosures vary widely and investors complain that they do not have the comparable, decision-useful information that they need to properly factor ESG considerations into their investment decisions. For their part, companies express concern over the lack of guidance as to which disclosure standards to follow. The SEC's reporting rules have remained essentially unchanged for decades even amid calls for reform. At the same time, organizations, including the World Economic Forum, have called for efforts to bring consistency and comparability to ESG disclosures. This chapter explores this dynamic landscape and anticipates significant change in the regulation of ESG reporting over the coming years.

Keywords ESG reporting · Stakeholder governance · Materiality · Disclosure provisions · Capital expenditures · Sustainability disclosures · Prescriptive disclosure rules · Voluntary disclosure standards · Transition risks · Physical risks · Convergence · Standardization

P. A. Davies · P. M. Dudek · K. S. Wyatt (✉)
Latham & Watkins LLP, Washington, D.C, USA
e-mail: Kristina.Wyatt@lw.com

© The Author(s) 2020
D. C. Esty and T. Cort (eds.), *Values at Work*,
https://doi.org/10.1007/978-3-030-55613-6_11

Environmental, social, and governance (ESG) issues have become a critical, mainstream global concern. In 2019, Time Magazine named climate change activist Greta Thunberg its Person of the Year[1] and protesters marched by the millions across the globe urging world leaders to address climate change.[2] The Oxford Dictionary named "Climate Emergency" as the 2019 Word of the Year after its use increased 100 fold over the prior year.[3] In 2019, CEOs of 181 companies across industries and sectors signed the Business Roundtable's Statement on the Purpose of a Corporation, which embraced "stakeholder governance," the proposition that corporations exist to serve a broad range of stakeholders, including their employees, customers, shareholders, and the broader communities in which they operate.[4]

As the focus on companies' impact in society has intensified, so too has investors' demand for information about companies' handling of their ESG risks and opportunities. In 2020, two of the world's largest asset managers, BlackRock and State Street, shook the corporate and investment world when they issued statements urging companies to improve their ESG performance or face divestment or board no-confidence votes.[5] The call for companies to address their impact on the environment and acknowledge their broader ESG risks and opportunities is loud and clear. Yet companies face significant challenges in determining which ESG information to disclose, and which standards to apply in crafting their disclosures. They also must balance investors' requests for additional information against the threat of litigation related to their disclosures.

ESG as a Mainstream Business Concern

The U.S. Securities and Exchange Commission (SEC) has provided only limited and high-level guidance as to which ESG information companies should disclose in their SEC filings. Investor groups have petitioned the SEC to enhance its disclosure requirements.[6] To date, the SEC has been reluctant to mandate additional disclosure of ESG matters, and companies therefore continue to have significant latitude to decide which ESG information to disclose and in which format. As a result, companies' ESG disclosures are widely varied. Some companies provide detailed metrics and analyses of their ESG factors, while other companies provide little or no ESG-focused disclosures.

In the absence of prescriptive guidance from the SEC, an array of voluntary reporting standards has emerged that seeks to establish the ESG information that companies should disclose. These different reporting standards serve a useful purpose in eliciting additional information from reporting companies, but their proliferation has also created uncertainty. The reporting landscape has become a patchwork of voluntary disclosure regimes and investor questionnaires that has left some issuers with questionnaire fatigue and others confused as to which guidance to follow and how to reconcile the different requests. Investors, for their part, continue to be dissatisfied with the information they receive, saying that disclosures are not decision-useful because they are neither consistent nor comparable from company to company.[7]

In 2020, the SEC's own Investor Advisory Committee urged the organization to adopt standardized ESG disclosure requirements, saying, "the time has come for the SEC to address this issue."[8] The Investor Advisory Committee underscored the importance of finding a unified approach to reporting, pronouncing,

> Investment and voting based in part on ESG disclosure is front and center in today's global investment ecosystem. Major business risks, decisions, and strategies stand upon ESG factors and investors are not being served or protected by the piecemeal, ad hoc, inconsistent information currently in the mix.[9]

A number of market participants have noted the divide between the information that companies are providing, on one hand, and the information that investors would find useful, on the other. The World Business Council for Sustainable Development conducted a series of investor round tables to identify the information that investors want in order to incorporate companies' sustainability performance in their investment processes.[10] The investors indicated that companies should discuss relevant risks clearly and with specificity. Further, investors want companies to demonstrate good governance and effective internal controls over ESG risks.[11] The participants noted the difficulty of incorporating non-financial information in their valuation models due to a lack of comparability across companies and a lack of narrative explanation to contextualize the metrics.[12]

The status quo is under immense criticism, given investor pressure for more consistent and actionable ESG information. The investment community can therefore expect to see continued strong advocacy for the adoption of more prescriptive reporting standards by the SEC, a convergence of voluntary ESG reporting standards to a more unified framework, or a combination of the two approaches.

The Reporting Framework in the United States

Overview of the SEC Reporting Requirements

The SEC's reporting rules and forms frame the information that companies include in the registration statements that they use to offer and sell securities and in the ongoing quarterly and annual reports that they issue to shareholders. Companies' disclosure documents contain two discrete types of information. Financial statements, prepared in accordance with Generally Accepted Accounting Principles, provide a numeric picture of company performance. Because the financial statements are prepared according to strict accounting rules and are audited by auditors, they provide a useful means by which investors can understand and compare companies.

Registration statements and periodic reports also include non-financial information, which provides context for the information in the financial statements and further detail about a company's operations and prospects. Some disclosure rules governing non-financial information are *prescriptive* and tell companies specifically which information must be disclosed. However, companies and their operations are widely varied, so crafting rules that define all the specific information that companies should include would be difficult. In recognition of this fact, many of the SEC's disclosure rules are *principles-based*, and articulate the purpose of the rule but leave it to the company to apply the rule to its own specific circumstances and to craft disclosures accordingly. This approach relies on the over-arching principle that companies should disclose to investors information that is *material* to them, that is, important to their decision to buy, sell, or vote a security.

The non-financial disclosure provisions of the SEC rules are designed to provide investors with a more complete picture of the company and its business. For example, companies must provide a narrative description of the business that the company is in, material legal proceedings that the company is subject to, material risk factors related to the company or the offering of securities, and a discussion and analysis of the company's financial condition and results of operations, compiled in the Management Discussion and Analysis disclosure ("MD&A"). The Management Discussion and Analysis disclosure provides management's discussion of factors that offer a fuller understanding of the company's business, including trends and uncertainties that could impact the business in the future.[13]

Application of the SEC's Disclosure Provisions to ESG Issues

The SEC has faced pressure for some time to adopt more stringent rules related to ESG disclosures. Its response has been to rely on its existing rules, which provide a framework for disclosure of material ESG issues. In 2010, as climate change was emerging as a business concern, the SEC issued guidance, the 2010 Interpretive Release, to help companies to apply the SEC's disclosure framework to climate change issues.[14] The SEC continues to point to the 2010 Interpretive Release, as the organization's Chairman Clayton did when discussing ESG disclosure principles in 2020.[15]

The 2010 Interpretive Release explained the various ways in which climate change issues could trigger disclosure obligations under the SEC's existing rules. The 2010 Interpretive Release does not reflect a new rule, but rather articulates the Commission's principles-based approach to its existing disclosure requirements which, based on assessments of materiality, can encompass new and evolving concerns. In the context of climate change, for example, a company would need to provide disclosure in its description of its business if compliance with environmental laws and regulations would have a material effect on its financial position, or if it needed to make material capital expenditures for environmental controls. Further, the SEC noted that disclosure might be required by companies making investments in new products or segments as a result of climate change. Climate change issues also could trigger disclosure in the Legal Proceedings or Risk Factors sections of a company's registration statements or reports where they present material risks or threats of litigation.

Disclosure could also be required in the Management Discussion and Analysis section of the registration statement or reports. The Management Discussion and Analysis disclosure provisions require companies to discuss their financial condition and liquidity, changes in financial condition, and the results of the company's operations. Companies must also describe any material known trends or uncertainties that could cause future results to differ from the company's reported results. The impacts of climate change, changes in customer demand for products, restrictions on supply due to supply chain disruptions, and other ESG factors might warrant Management Discussion and Analysis disclosure if management deems them material to the company.

The 2010 Interpretive Release also discussed how developments in federal, state, and local laws, rules, and regulations could trigger disclosure obligations under the SEC's existing disclosure rules. The release provided examples of regulatory developments that could have impacts that would require disclosure if material to the company. These developments might include the cost to purchase carbon allowances pursuant to cap-and-trade programs, the cost of improving facilities or equipment in order to comply with regulatory limits on emissions, and financial impacts from increased or decreased demand for goods due to developments such as carbon taxes.

Climate change can also have indirect effects that, if material, might trigger disclosure obligations. For example, customers' concern about climate change could increase or decrease demand for the company's products. Similarly, the SEC urged companies to consider the effects of reputational impacts related to climate change, including the risk of boycotts or protests that might materially impact the company.

The 2010 Interpretive Release stressed that the physical effects of climate change, such as flooding, hurricanes, rising sea levels, rising temperatures, or impaired access to water, could present threats to a company's operations that, if material, would require disclosure. The SEC noted that the increased frequency of severe weather could raise the risk of: property damage and disruption to operations, financial and operational impacts due to disruptions to key customers or suppliers, increased insurance claims for insurance companies and higher premiums for companies in coastal areas, and decreased agricultural production and capacity in areas impacted by flooding or drought.[16]

Pressure on the SEC to Amend Its Rules to Promote ESG Disclosures

Even though the SEC maintains that its existing disclosure rules already elicit disclosure of material ESG issues, many investors complain that companies are not providing meaningful disclosures with regard to their ESG risks. The point was punctuated by investors' responses to a 2016 Concept Release, in which the SEC solicited public comment on modernizing the disclosure requirements applicable to public companies.[17] This Concept Release specifically included questions about ESG disclosures, noting that "the role of sustainability and public policy information in investors' voting and investment decisions may be evolving as some investors are increasingly engaging on certain ESG matters."[18]

The Concept Release received a significant response, with comments disproportionately focused on sustainability disclosures. Of its 92 pages, only four were devoted to sustainability, yet a significant majority of the comment letters addressed sustainability issues generally, and climate change in particular.[19] In addition to climate change, commenters asked the SEC to address disclosures related to access to and stewardship of water, land tenure rights, diversity, gender pay equity, human rights, human capital management, sustainable palm oil, forestry, and supply-chain management.[20]

A common theme of the comments was that the quality and consistency of ESG disclosures need improvement. Keith Higgins, then the Director of the Division of Corporation Finance, summarized the point: "Many of the commenters found voluntary disclosures to be inconsistent, difficult to find, and often not comparable and lacking in context."[21]

Since the time of the Concept Release, the SEC has engaged in rule-making to update its disclosure rules. However, to the disappointment of many investors and investor advocates, these proposals have not answered the calls for enhanced ESG disclosure requirements.

In 2019, the SEC proposed rule amendments to update the provisions governing the Description of Business, Legal Proceedings, and Risk Factor disclosures.[22] The SEC's adherence to a principles-based approach to rulemaking was a recurring theme throughout the release. In 2020, the SEC proposed amendments to the rules underlying the disclosure provisions related to Management Discussion and Analysis.[23] This release sought to modernize and simplify the disclosure requirements, again applying a principles-based approach. This proposal, like the 2019 proposal, did not address the requests for enhanced ESG disclosure rules made by the commenters on the 2016 Concept Release.

On the day that the SEC issued the proposed Management Discussion and Analysis amendments, SEC Chairman Jay Clayton and Commissioner Allison Herren Lee issued separate statements addressing ESG disclosures. Chairman Clayton noted the complexity and difficulty of regulating ESG disclosures. He stressed that "the landscape around these issues is, and I expect will continue to be, complex, uncertain, multi-national/jurisdictional and dynamic." [24] The Chairman noted that he has been engaged in discussions with a variety of market participants as well as with his international counterparts on the issue of ESG disclosures.

Commissioner Lee issued her own statement expressing her disappointment that the SEC was not proposing amendments to the disclosure rules to address ESG issues, saying:

> Today's proposal is most notable for what it does not do: make any attempt
> to address investors' need for standardized disclosure on climate change risk...
> investors are overwhelmingly telling us, through comment letters and petitions
> for rulemaking, that they need consistent, reliable, and comparable disclosures

of the risks and opportunities related to sustainability measures, particularly climate risk.[25]

A number of factors might contribute to the SEC's reticence to consider more prescriptive disclosure rules around ESG reporting. To begin, ESG issues have only recently been broadly recognized as key economic issues that are material to mainstream investors. A key SEC mandate is to protect investors. When it adopts disclosure rules, it must undertake a cost-benefit analysis that weighs, among other factors, the costs to companies of providing disclosures against the benefits to investors. The utility of ESG information to mainstream investors has only become evident relatively recently.

Even now that ESG issues are recognized as material, regulating their disclosure in a prescriptive manner is not easy, particularly given the wide swath of issues encompassed under the ESG umbrella. Mandating disclosure of matters that are material to only some companies could lead to disclosure documents that are cluttered with extraneous information, potentially obfuscating the material information that investors truly need. This variability in the types of ESG information that is material to different companies might explain the SEC's continued adherence to a principles-based approach. One way to respond to that concern is to do as the SEC has done—to establish broad rules that articulate common disclosure principles and leave companies to determine how to apply their facts to those rules. Yet, as a result, some companies' disclosures are considered less rigorous and decision-useful than they might be under prescriptive disclosure rules.

Another factor that makes the regulation of ESG disclosures difficult is the forward-looking nature of much of the information that would be most useful to investors. Though some ESG information naturally is historical, such as historical greenhouse gas emissions, much ESG information that is most useful to investors requires forward-looking projections. Further, ESG information can be difficult to calculate, and involves inherent uncertainty. Companies are correctly concerned about the potential legal exposure they might face if their projections or forecasts of future impacts ultimately prove inaccurate. In addition to the forward-looking analysis provided in Management Discussion and Analysis, which includes material known trends and uncertainties that could cause future results to differ from those in the financial statements, the SEC principally regulates the disclosure of historical information.

Voluntary Disclosure Standards Seek to Fill the Breach

Given investors' and other stakeholders' dissatisfaction with the state of ESG disclosures, a host of "voluntary" disclosure standards has emerged to fill the informational gaps. Some of the more prominent standards are outlined below.

CDP

The CDP (formerly known as the Carbon Disclosure Project) operates a disclosure system that tracks the environmental impact of companies, municipalities, and other entities.[26] According to its website, the CDP has built the most comprehensive set of self-reported environmental data in the world, with more than 7000 companies and 620 cities reporting environmental data through the CDP in 2019.[27] The CDP requests detailed information on participants' environmental performance, greenhouse gas emissions, and environmental governance. The CDP then analyzes that data with reference to critical environmental risks and opportunities and shares the analyses and resulting scores with investors and other stakeholders with an interest in the information.

Climate Disclosure Standards Board

The Climate Disclosure Standards Board (CDSB) is another private sector organization focused on incorporating environmental effects in mainstream financial reporting by addressing the treatment of "natural capital" alongside financial capital. The Board is "committed to advancing and aligning the global mainstream corporate reporting model to equate natural capital with financial capital."[28] The CDSB's Climate Change Reporting Framework helps companies to report environmental information with a level of rigor comparable to that applied to financial information.

Global Reporting Initiative

The Global Reporting Initiative (GRI) was formed in 1997 to help companies and governments better understand and communicate their impact on sustainability issues such as climate change, human rights, governance, and social well-being.[29] According to the Initiative, "of the world's largest 250 corporations, 92% report on their sustainability performance and 74% of these use GRI's standards."[30]

Sustainability Accounting Standards Board

The Sustainability Accounting Standards Board (SASB) is a private sector body that was formed to help businesses to identify, manage, and report on the sustainability topics that are most important to investors.[31] The Sustainability Accounting Standards Board produced a set of 77 industry-specific standards that target the sustainability issues that generally are most important within an industry. The industry focus helps companies identify the issues most salient to their businesses, and can facilitate comparability of disclosures across companies within an industry.

Task Force on Climate-Related Financial Disclosures

The Financial Stability Board (FSB) formed the Task Force on Climate-related Financial Disclosures (TCFD) in 2015 in order to develop a consistent set of standards for companies to make voluntary climate-related financial disclosures for investors, lenders, and other stakeholders.[32] The Task Force's recommendations articulate four core themes: (1) disclosures should describe the organization's governance with regard to climate-related risks and opportunities; (2) disclosures should explain how climate-related risks and opportunities could impact the company's business, financial condition, and strategy; (3) disclosures should explain how the organization identifies, assesses, and manages climate-related risks, including through scenario analyses; and (4) disclosures should use metrics and targets to evaluate and manage these risks and opportunities.[33]

The TCFD explains that organizations might face climate-related transition risks and risks associated with the physical impacts of climate change. Transition risks might include policy and legal developments, technological improvements that displace old systems, market risks, and reputational risks associated with changing customer perceptions of the organization's business. Physical risks might include damage to property due to rising sea levels or extreme weather, in addition to resource scarcity and supply-chain risks. Companies might also find opportunities resulting from their climate change strategies, including opportunities around energy efficiency, resource reuse, and the development of new products and markets.

United Nations Sustainable Development Goals

As part of its 2030 Agenda for Sustainable Development, the United Nations adopted the 17 Sustainable Development Goals (SDGs) and 169 specific targets. The Goals "recognize that ending poverty and other deprivations must go hand-in-hand with strategies that improve health and education, reduce inequality, and spur economic growth—all while tackling climate change and working to preserve our oceans and forests."[34] As organizations map their activities to the UN SDGs, they are encouraged to establish targets that are suitable for their own circumstances and will advance progress on the selected SDGs.

Reconciling the Various Reporting Standards

The SEC's disclosure requirements are generally only the starting point in companies' assessment of which ESG information to disclose. Many companies also follow voluntary reporting standards such as those described above. This approach can help companies to provide more complete disclosure of ESG issues while still adhering to the SEC's disclosure rules. At the same time, the proliferation of voluntary reporting standards and the lack of consistency among them has left many companies wondering which standards to follow. Further, while individual companies' disclosures are becoming more robust, investors still struggle with wide variances in disclosures that make comparison across companies difficult.

Various initiatives have attempted to help market participants navigate and rationalize the different reporting standards. The World Business Council for Sustainable Development's Reporting Exchange is an online platform that aggregates nearly 2000 mandatory and voluntary ESG reporting standards and frameworks in 70 countries,[35] and the same organization's ESG Disclosure Handbook provides further guidance for companies as they approach their ESG reporting processes.[36] The Disclosure Handbook is designed to help companies navigate the disclosure process, giving consideration to the informational demands of multiple stakeholders and the array of reporting standards.[37]

The Corporate Reporting Dialogue also aims to rationalize the ESG reporting landscape.[38] Its participants include the CDP, Climate Disclosure Standards Board (CDSB), Global Reporting Initiative (GRI), Sustainability Accounting Standards Board (SASB), International Accounting Standards

Board, and Financial Accounting Standards Board. The Corporate Reporting Dialogue has endeavored to reconcile the different reporting regimes by providing comparisons and summaries of the principal reporting standards, including a "landscape map" that compares the member organizations' disclosure standards.[39] The goal of the Corporate Reporting Dialogue is "to promote greater coherence, consistency and comparability between corporate reporting frameworks, standards and related requirements."[40]

The Corporate Reporting Dialogue is a sponsor of the Better Alignment Project, which aims to map the key provisions of the CDP, Climate Disclosure Standards Board, Global Reporting Initiative, Sustainability Accounting Standards Board, International Integrated Reporting Council (IIRC), and Task Force on Climate-related Financial Disclosures (TCFD) to find points of intersection that can be harmonized.[41] Consistent with the objectives of the International Integrated Reporting Council, the Better Alignment Project aims to facilitate integrated disclosure of financial and non-financial information.

SASB and the CDSB also have published a TCFD Implementation Guide designed to help companies apply the TCFD recommendations in harmony with SASB and CDSB standards in order to improve companies' climate-related disclosures.[42]

The Future of ESG Reporting: A Necessary Drive Toward Convergence

The importance of convergence of ESG reporting standards is clear. During the early months of 2020, both the World Economic Forum (WEF) and the International Organization of Securities Commissions (IOSCO) issued important documents calling for a convergence of standards. In 2020, the International Business Council of the World Economic Forum issued a consultation draft that proposed a unified set of ESG reporting standards. It explained,

> The absence of a generally accepted international framework for the reporting of material aspects of ESG and other relevant considerations for long-term value creation contrasts with the well-established standards that exist for reporting and verifying financial performance. The existence of multiple ESG measurement and reporting frameworks and lack of consistency and comparability of metrics were identified as pain points that hinder the ability of companies to meaningfully and credibly demonstrate the progress they are making on sustainability, including their contribution to the SDGs.[43]

The World Economic Forum Consultation Draft proposes a standardized set of 22 "core metrics," and separate "expanded metrics" that can form the basis of a unified reporting structure. The proposed metrics do not attempt to create entirely new reporting standards but rather draw from existing reporting standards, including the Global Reporting Initiative (GRI), Task Force on Climate-related Financial Disclosures (TCFD), Sustainability Accounting Standards Board (SASB), CDP, and the UN SDGs.

The International Organization of Securities Commissions (IOSCO) issued its Final Report on "Sustainable Finance and the Role of Securities Regulators and IOSCO" in 2020.[44] The report found that three themes have emerged: "multiple and diverse sustainability frameworks and standards, including sustainability-related disclosure; a lack of common definitions of sustainable activities; and greenwashing and other challenges to investor protection."[45] In response, the International Organization of Securities Commissions has established a Board-Level Taskforce on Sustainable Finance to improve disclosures by companies and asset managers, and to work in collaboration with other regulators and international organizations to enhance consistency and coordination of regulatory efforts.

We expect that these efforts by the World Economic Forum and the International Organization of Securities Commissions, as well as the calls for convergence of ESG reporting standards by academics and others, will lead to a unified reporting structure of some form, as yet to be precisely defined.[46] We believe that such a unified structure should include certain elements, many of which appear in the proposals currently under consideration. These are:

- *Consistency and comparability of key metrics*: Certain key metrics common to all companies should be reported in a uniform manner across companies, industries, and jurisdictions in which the company trades. These key metrics facilitate comparison across companies and should reduce the questionnaires and inquiries to which companies are currently subjected.
- *Industry-specific disclosures*: The reporting framework should recognize that many ESG issues are industry-specific and require industry-focused reporting standards, similar to the standards of the Sustainability Accounting Standards Board.
- *Use of existing standards*: As the World Economic Forum Discussion Draft notes, it should not be necessary to invent a new set of standards out of whole cloth. Rather, the new unified standards should draw on the

existing standards under which companies have been reporting, to the extent possible.

- *Discussion and analysis*: The data reported should be supplemented with a narrative discussion, similar to the Management Discussion and Analysis, that provides management's view as to the significance of the ESG data to the company and its prospects. This analysis well might follow the Task Force on Climate-related Financial Disclosures (TCFD) recommendations, expanded, as appropriate, beyond climate-related disclosures.
- *Data quality*: The quality and comparability of the reported data will be critical to its usefulness. The methodologies for gathering and reporting ESG data will need to be standardized, and in all likelihood, some form of assurance will help to ensure consistency and comparability of data.
- *Liability*: ESG data is inherently speculative and the effects of matters such as climate change are difficult to predict with specificity. If companies are to be asked to provide more meaningful disclosures, they should have protection from liability for their projections, made in good faith. A unified standard should include a safe harbor provision that provides protection to companies that disclose forward-looking ESG data in good faith, based on the information currently available. The specific contours of such a safe harbor would need to be determined, and could include conditions, such as the articulation of factors that could cause the actual results to differ. It also could be conditioned on the company having obtained partial or reasonable assurance of the ESG disclosures or other form of independent review such as the second party opinions provided in the context of green bonds.[47] Safe harbor protection might also be available to companies that have applied an accepted methodology for verifying and delivering the projections.

Conclusion

The ESG reporting landscape is dynamic, fragmented, and evolving. Companies operate in an environment in which the SEC reporting rules have remained essentially unchanged for decades even as investors complain that the ESG information they currently receive is inconsistent across companies and is not sufficiently helpful to their investment analyses. Even in the face of this investor pressure, the SEC has held to its long-standing application of a largely principles-based approach to the regulation of ESG disclosures. The SEC Chairman has indicated that the Commission continues to closely monitor ESG issues and to listen to its various constituents. Outside of the

SEC's rules, issuers increasingly apply voluntary standards to shape their ESG disclosures. Investors also have attempted to fill the informational gaps by issuing questionnaires to companies seeking further ESG data. Further, ESG surveys, ratings, and rankings have proliferated to meet investors' informational needs. The landscape remains crowded, confusing, and marked with some dissatisfaction on the parts of both investors and companies. This disclosure arena is changing, and will require close attention over the coming months and years as regulatory requirements, and guidance take shape, and as disclosure practices evolve.

Against the backdrop of this landscape are calls for reform. Most notably, the World Economic Forum and the International Organization of Securities Commissions are both calling for a convergence of standards. If we are to reach a point at which investors have the tools to compare the ESG performance of companies, a common reporting standard must be developed that can serve as the unified ESG reporting standard and address the current reporting fragmentation and confusion. We expect to see convergence on such a standard in the coming years with some concomitant movement by the SEC toward enhanced ESG disclosure rules, consistent with the new standards.

Notes

1. Alter, C., Haynes, S., & Worland, J. (2019). Time 2019 Person of the Year—Greta Thunberg. *Time*. Retrieved from https://time.com/person-of-the-year-2019-greta-thunberg/.
2. Taylor, M., Watts, J., & Bartlett, J. (2019, September 27). Climate Crisis: 6 Million People Join Latest Wave of Global Protests. *The Guardian*. Retrieved from https://www.theguardian.com/environment/2019/sep/27/climate-crisis-6-million-people-join-latest-wave-of-worldwide-protests.
3. Zhou, N. (2019, November 20). Oxford Dictionaries Declares 'Climate Emergency' the Word of 2019. *The Guardian*. Retrieved from https://www.theguardian.com/environment/2019/nov/21/oxford-dictionaries-declares-climate-emergency-the-word-of-2019.
4. Business Roundtable. (2019, August 19). Business Roundtable Redefines the Purpose of a Corporation to Promote "an Economy That Serves All Americans." *Business Roundtable*. Retrieved from https://www.businessroundtable.org/business-roundtable-redefines-the-purpose-of-a-corporation-to-promote-an-economy-that-serves-all-americans.
5. Fink, L. (2020). A Fundamental Reshaping of Finance. *BlackRock*. Retrieved from https://www.blackrock.com/corporate/investor-relations/larry-

fink-ceo-letter; Chasan, E. (2020, January 28). *State Street Tells Companies That ESG Moves Are No Longer Optional.* Bloomberg Green. Retrieved from https://www.bloomberg.com/news/articles/2020-01-28/state-str eet-tells-companies-esg-moves-are-no-longer-optional.

6. See, e.g., Williams, C. A. (2018, October 1). *Petition for Rulemaking on Environmental, Social, and Governance (ESG) Disclosure.* Retrieved from https://www.sec.gov/rules/petitions/2018/petn4-730.pdf. In addition, several ESG disclosure bills have been introduced in the US Congress. See, e.g., the Climate Risk Disclosure Act of 2019 (approved by the House Financial Services Committee), HR3623—116th Congress (2019–2020).

7. In response to this felt need for a common set of reporting standards, the World Economic Forum announced an effort in January 2020 to develop a core set of ESG disclosures. Cann, O. (2020, January 22). Measuring Stakeholder Capitalism: World's Largest Companies Support Developing Core Set of Universal ESG Disclosures. *World Economic Forum.* Retrieved from https:// www.weforum.org/press/2020/01/measuring-stakeholder-capitalism-world-s-lar gest-companies-support-developing-core-set-of-universal-esg-disclosures/.

8. Bennington, A. (2020, May 28). Recommendation from the Investor-as-Owner Subcommittee of the SEC Investor Advisory Committee Relating to ESG Disclosure. *Harvard Law School Forum on Corporate Governance.* Retrieved from https://corpgov.law.harvard.edu/2020/05/28/recommendation-from-the-investor-as-owner-subcommittee-of-the-sec-investor-advisory-commit tee-relating-to-esg-disclosure/.

9. Bennington, A. (2020, May 28). Recommendation from the Investor-as-Owner Subcommittee of the SEC Investor Advisory Committee Relating to ESG Disclosure. *Harvard Law School Forum on Corporate Governance.*

10. World Business Council for Sustainable Development. (2018). *Enhancing the Credibility of Non-financial Information: The Investor Perspective.* Retrieved from https://docs.wbcsd.org/2018/10/WBCSD_Enhancing_Credibility_Report.pdf.

11. World Business Council for Sustainable Development. (2018). *Enhancing the Credibility of Non-financial Information: The Investor Perspective.*

12. World Business Council for Sustainable Development. (2018). *Enhancing the Credibility of Non-financial Information: The Investor Perspective* (p. 7).

13. The non-financial disclosure provisions are set forth in SEC Regulation S-K, found at 17 CFR §229.

14. Commission Guidance Regarding Disclosure Related to Climate Change, Release Nos. 33-9106; 34-61469; 17 CFR Parts 211, 231, 241 (February 8, 2010).

15. Clayton, J. (2020, January 30). Statement on Proposed Amendments to Modernize and Enhance Financial Disclosures; Other Ongoing Disclosure Modernization Initiatives; Impact of the Coronavirus; Environmental and Climate-Related Disclosure. *U.S. Securities and Exchange Commission.* Retrieved from https://www.sec.gov/news/public-statement/clayton-mda-2020-01-30.

16. Clayton, J. (2020, January 30). Statement on Proposed Amendments to Modernize and Enhance Financial Disclosures; Other Ongoing Disclosure Modernization Initiatives; Impact of the Coronavirus; Environmental and Climate-Related Disclosure. *U.S. Securities and Exchange Commission* (p. 27).

17. SEC Concept Release, "Business and Financial Disclosure Required by Regulation S-K," Release No. 33-10064; 34-77599 (April 13, 2016). The Concept Release was part of the SEC's multi-faceted years-long Disclosure Effectiveness Initiative that is designed to improve the disclosure regime for both companies and investors by evaluating which information is disclosed, and how and where such disclosures are presented.

18. Ibid, at 211.

19. Sustainability Accounting Standards Board. (n.d.). *Business and Financial Disclosure Required by Regulation S-K—The SEC's Concept Release and Its Implications*. Retrieved from https://www.sasb.org/wp-content/uploads/2016/09/Reg-SK-Comment-Bulletin-091416.pdf.

20. The SASB summary provided that the SASB itself has not determined that all of these issues likely encompass material information across all industries and therefore are not all included in the SASB disclosure framework. See SASB analysis, *supra* note 105, at footnote 11.

21. Ibid, at 3.

22. SEC Proposed Rule, "Modernization of Regulation S-K Items 101, 103, and 105," SEC Release Nos. 33-10668, 34-86614 (August 8, 2019).

23. SEC Proposed Rule, "Management's Discussion and Analysis, Selected Financial Data, and Supplementary Financial Information," SEC Release Nos. 33-10750; 34-88093 (January 30, 2020).

24. Clayton, J. (2020, January 30). Statement on Proposed Amendments to Modernize and Enhance Financial Disclosures; Other Ongoing Disclosure Modernization Initiatives; Impact of the Coronavirus; Environmental and Climate-Related Disclosure. *U.S. Securities and Exchange Commission*.

25. Lee, A. H. (2020, January 30). "Modernizing" Regulation S-K: Ignoring the Elephant in the Room. *U.S. Securities and Exchange Commission*. Retrieved from https://www.sec.gov/news/public-statement/lee-mda-2020-01-30.

26. CDP. (2020). *CDP.* Retrieved from https://www.cdp.net.

27. Ibid.

28. Ibid.

29. Global Reporting Initiative. (2020). About GRI. *GRI.* Retrieved from https://www.globalreporting.org/Information/about-gri/Pages/default.aspx.

30. Global Reporting Initiative. (2020). GRI and Sustainability Reporting. *GRI.* Retrieved from https://www.globalreporting.org/information/sustainability-reporting/Pages/gri-standards.aspx.

31. Sustainability Accounting Standards Board. (2018). *SASB.* Retrieved from https://www.sasb.org/.

32. Task Force on Climate-related Financial Disclosures. (2020). About the Task Force. *TCFD.* Retrieved from https://www.fsb-tcfd.org/about/.

33. Ibid.
34. United Nations. (2015). Sustainable Development Goals. *Sustainable Development Goals Knowledge Platform*. Retrieved from https://sustainabledevelopment. un.org/?menu=1300.
35. World Business Council for Sustainable Development. (2020). The Reporting Exchange. *WBCSD*. Retrieved from https://www.wbcsd.org/Programs/Redefi ning-Value/External-Disclosure/The-Reporting-Exchange.
36. World Business Council for Sustainable Development. (2019, April 3). ESG Disclosure Handbook. *WBCSD*. Retrieved from https://www.wbcsd.org/Pro grams/Redefining-Value/External-Disclosure/Purpose-driven-disclosure/Resour ces/ESG-Disclosure-Handbook.
37. World Business Council for Sustainable Development. (2019, April 3). ESG Disclosure Handbook. *WBCSD*.
38. International Integrated Reporting Council. (2020). *Corporate Reporting Dialogue*. Retrieved from https://corporatereportingdialogue.com/.
39. International Integrated Reporting Council. (2020). The Landscape Map. *Corporate Reporting Dialogue*. Retrieved from https://corporatereportingdia logue.com/landscape-map/.
40. International Integrated Reporting Council. (2020). About. *Corporate Reporting Dialogue*. Retrieved from https://corporatereportingdialogue.com/about/.
41. International Integrated Reporting Council. (2020). Better Alignment Project. *Corporate Reporting Dialogue*. Retrieved from https://corporatereportingdia logue.com/better-alignment-project/.
42. Sustainability Accounting Standards Board & Climate Disclosures Standards Board. (2019). *TCFD Implementation Guide: Using SASB Standards and the CDSB Framework to Enhance Climate-Related Financial Disclosures in Mainstream Reporting*. Retrieved from https://www.sasb.org/kno wledge-hub/tcfd-implementation-guide/; Davies, P. A., & Wyatt, K. S. (2019, May 13). SASB and CDSB Issue TCFD Implementation Guide. *Latham & Watkins LLP Environment, Land & Resources Blog*. Retrieved from https://www.globalelr.com/2019/05/tcfd-issues-implementation-guide-inc orporating-sasb-and-cdsb-frameworks/.
43. World Economic Forum. (2020). *Toward Common Metrics and Consistent Reporting of Sustainable Value Creation*. Retrieved from http://www3.weforum. org/docs/WEF_IBC_ESG_Metrics_Discussion_Paper.pdf.
44. The Board of the International Organization of Securities Commissions. (2020). *Sustainable Finance and the Role of Securities Regulators and IOSCO*. Retrieved from https://www.iosco.org/library/pubdocs/pdf/IOSCOPD652.pdf.
45. The Board of the International Organization of Securities Commissions. (2020). *Sustainable Finance and the Role of Securities Regulators and IOSCO*.
46. Daniel C. Esty, Todd Cort, et al. "Toward Enhanced Sustainability Disclosure: Identifying Obstacles to Broader and More Actionable ESG Reporting," forthcoming White Paper.

47. Companies currently can avail themselves of the protections of the safe harbor in the Private Securities Litigation Reform Act in the United States. However, that safe harbor only applies to private litigation and only for companies reporting in the United States. Further, the applicability of the safe harbor in the context of ESG disclosures has not been tested.

12

Reasonable Investors' Growing Awareness of Climate Risk and Its Impact on U.S. Corporate Disclosure Law

Hana V. Vizcarra

Abstract Climate change impacts have become more evident, triggering concern and calls to action among individuals, governments, and the private sector alike. Shareholders increasingly view information about the risk climate change poses to individual companies as critical to their investment decision-making. Their use of this information in turn influences corporate disclosure practices. Due to the quirks of a legal standard in the United States that evolves as investors' views change, what climate-change-related information companies need to include in their disclosures may shift even without any significant change to existing regulations. Indeed, there is evidence that this transformation has already begun. As this happens, courts will help define the parameters of what climate-change-related information is deemed *material*. This chapter discusses the trends in investor treatment of climate-related information, how it could impact corporate disclosure requirements, and how courts may view the application of U.S. securities law to climate change information.

Keywords Reasonable investor · Institutional investors · Corporate disclosures · Materiality · ESG issues · Securities law · Pension funds · Total mix of information

H. V. Vizcarra (✉)
Harvard Law School Environmental & Energy Law Program (EELP),
Cambridge, MA, USA
e-mail: hvizcarra@law.harvard.edu

© The Author(s) 2020
D. C. Esty and T. Cort (eds.), *Values at Work*,
https://doi.org/10.1007/978-3-030-55613-6_12

Investors have awakened to the physical and transitional risks of climate change. As institutional investors pay more attention to climate change, they increase pressure on companies to more deeply consider and disclose climate change risks and opportunities. This drumbeat for more expansive climate-related corporate disclosures has grown louder and more consistent, sparking a dialogue about the inadequacy and future of corporate disclosures.

This mounting pressure matters for the law. Calls for new mandates, laws, and guidance on corporate disclosure of climate-related information have emerged, but U.S. regulators have yet to substantially address these calls. Some international jurisdictions have started down the path of mandating disclosure of a broad array of environmental, social, and governance information, including climate change-related risk. The United States has not yet made such definitive moves, but that does not mean the status quo remains in place. As investors integrate climate-related data into their analysis, they shift what information is considered *material* under U.S. security law. This shift can change what information companies must disclose to federal regulators and shareholders even in the absence of significant regulatory or legislative changes.

This chapter discusses the trends that indicate a growing need to consider climate-change-related information material, how this information may be interpreted by the courts, and why that matters to continued efforts to encourage more expansive corporate disclosures related to climate change.

Importance of Analyzing Existing Law in Addition to Imagining Future Reforms

Myriad articles discuss corporate responsibility and environmental, social, and governance (ESG) issues across disciplines.[1] U.S. securities law and its disclosure regime, including the meaning of materiality as defined by the Supreme Court and the Securities and Exchange Commission (SEC), have likewise been the subject of much discussion.[2] While recent papers have considered proposals for incorporating ESG issues into legally mandated disclosures, fewer have considered how courts view the materiality of ESG and, more specifically, climate-related information.[3]

Federal regulators, particularly the SEC, have not adequately addressed the rapidly changing climate discussion, the rise of its importance to investors, or the certainty of climate-related impacts. The SEC has not provided guidance on the materiality of climate change-related disclosures since 2010.[4] It has largely remained on the sidelines since then, leaving companies and investors to spar over how expansive climate-related disclosures should be, on what topics, and in what forms. In May 2020, a subcommittee of the SEC's Investor Advisor Committee recommended that the SEC incorporate material ESG factors into reporting requirements.[5] Even with this recommendation, we are unlikely to see new guidance, enforcement activity, or legislative action clarifying what type of climate-related information fits into the existing disclosure regime in the very near term.

Courts may consider what climate-related information companies must disclose before any legislative and regulatory changes appear. In doing so, courts will assess the evolution of what climate-related information the *reasonable investor* considers *material* to their investment decisions. Securities law requires companies to disclose certain *material* information to investors and imposes liability for omissions and untrue or misleading statements. Court decisions will guide and potentially limit any efforts by the SEC to update disclosure requirements to specifically account for climate change information. Understanding how a court might apply the definition of materiality to climate-related information provides insight into possible future regulatory requirements and a preview of what could become required disclosures even absent new regulation. Recent trends inform how a court will view the materiality of climate-related information.

Trends Indicate the Materiality of Climate Change-Related Information Is Growing

Four trends in the corporate-investor disclosure dance indicate that today's reasonable investor considers climate-related information material: (1) the growing interest by mainstream investors in climate-related information; (2) investors' use of the climate information they get from companies; (3) companies' responsiveness to investor demands for more information; and (4) the increasing importance of the views of the largest fund managers on disclosures.

Investors Show Interest in Climate Change Information

A growing number of investors—beyond impact investors—have made clear they see climate-related information as potentially material and want that information reflected in corporate disclosures. Some have adopted the Principles for Responsible Investment, designed with the support of the United

Nations to help incorporate ESG factors into investment and ownership decisions. Other investors have supported the G-20 Financial Stability Board's Task Force on Climate-related Financial Disclosures (TCFD). Leading investors have given speeches on the "tragedy of the horizon"[6] and issued reports, letters, and memoranda touting the importance of climate information. Mainstream investors, voluntary reporting organizations, and ratings agencies support the TCFD's recommendations and have sought detailed, expansive, and data-supported information. Major asset managers have voted in support of efforts to improve corporate governance on climate change, and pension funds have made commitments on disclosure and climate-related investments.[7]

Investors Use Climate-Related Information

One key to pinpointing the reasonable investor's view on the materiality of climate-related information is how investors are using the disclosed information. Courts considering whether information in any given case is material rely on such concrete evidence of its importance to investors. Surveys show investors consider ESG information when making investment decisions because they believe it is their fiduciary duty and financially material to investment performance.[8] Investors actively address climate change with companies, using engagement, shareholder proposals, and the power of their proxy votes to encourage disclosure of climate-related information.

High-profile announcements in early 2020 by State Street Global Advisors, BlackRock, and other investment firms outlined specific plans for persuading companies to address financially material ESG issues. BlackRock outlined changes to its investment management practices to integrate climate-related risk, called for improved corporate disclosures in accordance with guidelines set forth by the Sustainable Accounting Standards Board (SASB) and the TCFD, and said it would hold accountable companies that do not effectively address material climate-related risks.[9] State Street Global Advisors said it would vote against boards of companies who fall behind in addressing financially material ESG issues.[10] This follows its 2019 creation of an ESG scoring system, the "responsibility factor," based on SASB's disclosure guidance.[11]

These announcements followed moves in recent years by Wellington Management, the California Public Employees' Retirement System, and other institutional investors to improve their capacity to consider climate change risk in their investment decisions.[12] Wellington Management and Woods Hole Research Center launched an initiative in 2018 to integrate climate science into Wellington's asset management by creating models to analyze climate change impacts on global capital markets, and the California Public Employees' Retirement System committed to applying the resulting insights in its portfolio. BlackRock partnered with Rhodium Group to identify how physical climate change risks impact financial performance. Investor advisor companies and ratings agencies have acquired climate data and risk analysis companies, responding to the demand from investors for better tools to assess these issues. In 2019, Moody's acquired climate data and risk analysis company Four Twenty Seven, Inc., and MSCI acquired Carbon Delta. S&P Global Ratings launched the ESG Evaluation program and ESG Risk Atlas designed to inform investors and companies of risks, including that of climate change. In 2017, Institutional Shareholder Services acquired the investment climate data division of the South Pole Group. These concrete efforts to incorporate climate change data and disclosures into investment management show that investors' interests extend beyond mere talk.

Corporate Response

Companies have responded to the rising importance of climate-related information among investors with a steady stream of climate-change-related goals, commitments to reporting in line with the TCFD recommendations, and issuance of climate change reports in addition to sustainability and annual reports. The number of companies disclosing ESG data has dramatically increased since the early 1990s.[13] As of 2019, 785 firms have committed to supporting the TCFD's recommendations, and top oil and gas companies have released special climate change reports.[14]

With the SEC's relative silence on the topic, companies have sought guidance on reporting from independent organizations and engagement with shareholders. They have relied on SASB's industry-specific guides and the TCFD's framework recommendations in deciding how to approach climate-change-related disclosures and continue to use other organizations' disclosure platforms and guidance on specific metrics.

These new disclosure practices, even where only apparent in voluntary disclosures outside of SEC filings, indicate a recognition by boards of the reasonableness of investor requests for more substantive climate-related disclosures.

Institutional Investor Consolidation

The rise of index funds like BlackRock, Vanguard, or State Street enables these organizations to increasingly control "how the rights associated with those shares are used for governance purposes."[15] In this context, these institutions' positions on climate-related disclosure play an outsized role in influencing corporate actions. The "Big Three" fund managers have all supported the TCFD recommendations and pressed for more disclosure of climate-change-related information. Such support may represent a concern for portfolio-level rather than firm-level profit maximization, but in either case it represents a rational interest.[16] Regardless of the primary motivating factor, courts can no longer view climate-related information as relegated to a niche subset of investors, dismissible out of hand. As these entities rely on the existence or absence of climate change information in making decisions, their positions on climate-related disclosure may be considered representative of the *reasonable investor*.

How Courts Might Interpret These Trends

The reasonable investors' evolving view of climate-related information means companies must evolve their approaches to making materiality determinations as well. Material information has a particular, if nebulous, definition in domestic securities law. The Supreme Court defined it in *TSC Industries, Inv. V. Northway, Inc.* as information for which there is "a substantial likelihood that a reasonable shareholder would consider it important in deciding how to vote," meaning its disclosure would have "significantly altered the 'total mix' of information made available."[17] This definition, adopted by regulators,[18] guides corporate financial reporting to the Securities and Exchange Commission and communications with shareholders. Companies must share certain information with investors and risk liability for making untrue statements, misleading investors, and omitting financially material information. Management and boards decide what to disclose,[19] but the definition of materiality requires them to consider the shareholder's viewpoint.

When, how, and even whether certain topics become material depends on case specifics.[20] When a court considers the materiality of a piece of information it does so in relation to the company's individual situation and how a reasonable investor would view it. There is no bright-line rule,[21] but there are also no "'degrees' of materiality"—information is either material or not.[22] This binary approach makes courts understandably wary of setting the threshold too low.[23] Courts contend that the reasonable investor standard is objective, measured by the views of the mainstream market as a whole, in which the reasonable investor is neither the worst nor best informed.[24] A reasonable investor must exercise care in considering information and take into account publicly available information and relevant industry customs,[25] but is not an expert. Despite prominence in the definition of materiality, investors do not make companies' disclosure decisions,[26] making court review of corporate disclosures all the more important.

Defining the reasonable investor in relation to the whole of investors engaged in the market allows for variability over time.[27] Federal regulators have recognized the potential for such a shift.[28] As more reasonable investors consider information about climate change risks and opportunities material to the total mix of information, the likelihood increases that courts will too.

It remains to be seen how the spike in investor focus on climate concerns will shape courts' understanding of the reasonable investor's expectations. Courts' evaluation of the question will impact how effectively climate-related information is integrated into mainstream investing. In cases involving environmental information, materiality findings generally coincide with acute events, such as spills or accidents.[29] Courts have also found substantial noncompliance with environmental regulations material.[30] Only two cases directly addressing climate-related disclosures have resulted in opinions, both involving the same basic facts and both involving information related to energy transition risks. These cases—one brought by shareholders and one by a state attorney general—provided the first opportunities for courts to opine on the materiality of certain climate-related information.

The first case to make it to the courtroom was a shareholder suit against Exxon Mobil Corporation.[31] The *Ramirez v. Exxon Mobil Corp.* opinion notably acknowledged that information representing climate change risks could be material to reasonable investors. However, this case only considered the question in the context of a motion to dismiss and did not review the merits of the arguments in full.

The New York attorney general brought the second case resulting in an opinion. In *New York v. Exxon Mobil Corp.*, the court considered whether Exxon Mobil misled investors in its disclosures about the potential impacts of future climate change policies on product demand and how it incorporated this information into its project-level business planning.[32] Plaintiffs failed to convince the court of the materiality of company's statements or supposed omissions.[33] The court found plaintiffs' experts unpersuasive and found no evidence of impact on investment analysts' analyses or actual investors' decisions during the relevant time frame.

The discussions of how to treat climate-related information in these cases may shape corporate materiality determinations in the near future but do not provide a clear path for how the law will develop. These cases acknowledged the potential materiality of climate-related information but did not ultimately find future cost estimates of an energy transition material to a reasonable investor's decisions. They also provide no insight into how courts might treat information about a company's exposure to the physical risks of climate change.

The *New York v. Exxon Mobil Corp.* opinion in particular deserves a closer look. It indicates companies have significant leeway in how they consider future transition risks as long as discussions of their evaluation and incorporation of those risks are not misleading. Of particular importance to other securities cases is how the judge discussed the way a reasonable investor would view cost assumptions that feed into modeling and projections for future costs and demand. The court stated that "[n]o reasonable investor during the period from 2013 to 2016 would make investment decisions based on speculative assumptions of costs that may be incurred 20+ or 30+ years in the future with respect to unidentified future projects."[34] The judge also found that the disclosures in the documents themselves were not misleading, that no actual investors were misled, and that the information in question did not impact investors' analysis of the company or its stock.

The finding that "no reasonable investor" would make investment decisions in the near-term based on costs projections 20+ years out may not remain true in a different context. Investors have recently shown interest in actively evaluating companies' views of potential future demand and costs, calling for more disclosure about how companies make these evaluations. And while investors may not make decisions based on future scenario projections alone, they might make decisions based on how well companies are prepared to adjust to possible futures and whether companies make a good faith effort to grapple with plausible scenarios. As investors start to incorporate these types of disclosures into new methods of analysis, this could change the "total mix" of information available to them.

Future of Materiality and the Reasonable Investor

The investor relationship to climate-related information has shifted even in the few years since the time frame addressed in the *New York v. Exxon Mobil Corp.* case. Investors are finding new ways to incorporate such information into their portfolio management practices. The court cases discussed above dealt with information about transition risk like modeling of future drops in demand and price shifts from a transition to a lower-carbon economy. While we have less information about how a court might view the materiality of information related to the physical risks of climate change, such information could cross the materiality threshold even sooner for some industries. The collaborations that have emerged between investors and climate data firms indicate investors are tuned into physical risks as well. As evidence of investors considering both physical and transition risk information grows, the support for and probability of a court finding such information *material* also grows. By doing so, a court would be confirming that *reasonable* investors would expect to see such information in disclosures.

The trends described highlight the importance for companies to clearly explain how they evaluate and consider climate-related information in a manner that does not risk misleading investors. Consensus around how to do so is slowly forming, with an emphasis on using the Task Force on Climate-related Financial Disclosures' framework and the Sustainable Accounting Standards Board's industry-specific guidance. There remain no bright lines between the important and the *material*, but as investors develop increasingly sophisticated efforts to incorporate climate-related information into their decision-making, the small drip of cases questioning the adequacy of corporate disclosures may become a torrent if companies do not more effectively address questions of climate impacts on their business.

This chapter was adapted from the author's piece titled "The Reasonable Investor and Climate-Related Information: Changing Expectations for Financial Disclosures" published in the February 2020 edition of the Environmental Law Reporter, an early version of which was presented at the November 2019 Yale Initiative on Sustainable Finance's Symposium. It builds on the author's 2019 article in the Vermont Law Review (cited in endnote 7). A related essay is being considered for publication in Fall of 2020 by the Environmental Law Forum.

Notes

1. Williams, C. A. (2018). Corporate Social Responsibility and Corporate Governance. In J. N. Gordon & W. G. Ringe (Eds.). *The Oxford Handbook of Corporate Law and Governance*. Oxford, UK: Oxford University Press.
2. Vizcarra, H. (2020). The Reasonable Investor and Climate-Related Information: Changing Expectations for Financial Disclosures. *Environmental Law*

Reporter, 50(2), 10106–10114. Retrieved from http://eelp.law.harvard.edu/wp-content/uploads/50.10106.pdf.

3. Vizcarra, H. (2020). The Reasonable Investor and Climate-Related Information: Changing Expectations for Financial Disclosures. *Environmental Law Reporter, 50*(2), 10106.

4. Vizcarra, H. (2020). The Reasonable Investor and Climate-Related Information: Changing Expectations for Financial Disclosures. *Environmental Law Reporter, 50*(2), 10111–10112.

5. Investor-as-Owner Subcommittee of the SEC Investor Advisory Committee. (2020). *Recommendation from the Investor-as-Owner Subcommittee of the SEC Investor Advisory Committee Relating to ESG Disclosure.* Retrieved from https://www.sec.gov/spotlight/investor-advisory-committee-2012/recommendation-of-the-investor-as-owner-subcommittee-on-esg-disclosure.pdf.

6. Carney, M. (2015). *Breaking the Tragedy of the Horizon—Climate Change and Financial Stability.* Retrieved from https://www.bis.org/review/r151009a.pdf.

7. Vizcarra, H. (2019). Climate-Related Disclosure and Litigation Risk in the Oil & Gas Industry: Will State Attorneys General Investigations Impede the Drive for More Expansive Disclosures? *Vermont Law Review, 43*(4), 733–775. Retrieved from http://eelp.law.harvard.edu/wp-content/uploads/VLR_Bk4_Vizcarra.pdf.

8. Kumar, R., Wallace, N., & Funk, C. (2020). *Into the Mainstream: ESG at the Tipping Point.* Harvard Law Forum on Corporate Governance. Retrieved from https://corpgov.law.harvard.edu/2020/01/13/into-the-mainstream-esg-at-the-tipping-point/; Amel-Zadeh, A. & Serafeim, G. (2018). Why and How Investors Use ESG Information: Evidence from a Global Survey. *Financial Analysts Journal, 74*(3), 87–103. Retrieved from https://www.cfainstitute.org/en/research/financial-analysts-journal/2018/faj-v74-n3-2.

9. Sorkin, A.R. (2020, January 14). BlackRock C.E.O. Larry Rink: Climate Crisis Will Reshape Finance. *New York Times.* Retrieved from https://www.nytimes.com/2020/01/14/business/dealbook/larry-fink-blackrock-climate-change.html; Fink, L. (2020). A Fundamental Reshaping of Finance. *BlackRock.* Retrieved from https://www.blackrock.com/corporate/investor-relations/larry-fink-ceo-letter; Fink, L. (2020). Sustainability as BlackRock's New Standard for Investing. *BlackRock.* Retrieved from https://www.blackrock.com/corporate/investor-relations/blackrock-client-letter.

10. Taraporevala, C. (2020, January 28). *CEO's Letter on SSGA 2020 Proxy Voting Agenda.* Retrieved from https://www.ssga.com/library-content/pdfs/insights/CEOs-letter-on-SSGA-2020-proxy-voting-agenda.pdf.

11. State Street Corporation. (2019). *R-Factor™: Reinventing ESG Investing Through a Transparent Scoring System.* Retrieved from https://www.ssga.com/investment-topics/environmental-social-governance/2019/04/inst-r-factor-reinventing-esg-through-scoring-system.pdf.

12. Vizcarra, H. (2020). The Reasonable Investor and Climate-Related Information: Changing Expectations for Financial Disclosures. *Environmental Law*

Reporter, 50(2), 10110 (describing and providing citations for the examples mentioned in this paragraph).

13. Amel-Zadeh, A. & Serafeim, G. (2018). Why and How Investors Use ESG Information: Evidence from a Global Survey. *Financial Analysts Journal, 74*(3), 87–103.

14. Task Force on Climate-Related Financial Disclosures. (2019). *2019 Status Report*. Retrieved from https://www.fsb-tcfd.org/wp-content/uploads/2019/06/ 2019-TCFD-Status-Report-FINAL-053119.pdf; Vizcarra, H. (2019). *Shifting Perspectives: E&P Companies Talk Climate and the Energy Transition*. Harvard Environmental & Energy Law Program. Retrieved from https://eelp.law.har vard.edu/2019/03/shifting-perspectives-ep-companies-talking-climate-and-the- energy-transition-trends-in-disclosure-and-climate-strategy/.

15. Coates, J. C. IV (2018). *The Future of Corporate Governance Part I: The Problem of Twelve* (Working Paper No. 19-07). Harvard Public Law. Retrieved from https://papers.ssrn.com/sol3/papers.cfm?abstract_id=3247337.

16. Condon, M. (2019). *Externalities and the Common Owner* (Law and Economics Research Paper No. 19-07). New York University School of Law. Retrieved from https://papers.ssrn.com/sol3/papers.cfm?abstract_id=3378783.

17. TSC Indus., Inc. v. Northway, Inc., 426 U.S. 438 (1976).

18. Vizcarra, H. (2019). Climate-Related Disclosure and Litigation Risk in the Oil & Gas Industry: Will State Attorneys General Investigations Impede the Drive for More Expansive Disclosures? *Vermont Law Review, 43*(4), 733–775. *See* Vizcarra, *supra* note 8, at 750 (noting the SEC adjusted its definition to align with the Supreme Court in Rule 12b-2; and citing Business and Financial Disclosure Required by Regulation S-K, Concept Release, 81 Fed. Reg. 23916, 23925 [April 22, 2016]).

19. Eccles, R. G., & Youmans, T. (2016). Materiality in Corporate Governance: The Statement of Significant Audiences and Materiality. *Journal of Applied Corporate Finance, 28*(2), 39–46. Retrieved from https://onlinelibrary.wiley. com/doi/epdf/10.1111/jacf.12173.

20. Vizcarra, H. (2019). Climate-Related Disclosure and Litigation Risk in the Oil & Gas Industry: Will State Attorneys General Investigations Impede the Drive for More Expansive Disclosures? *Vermont Law Review, 43*(4), 733–775. See Vizcarra, *supra* note 8, at 751 (citing SEC Staff Accounting Bulletin No. 99, 64 Fed. Reg. 45150, 45151 [August 19, 1999] [recommending consideration of qualitative factors and analysis of all relevant considerations when determining materiality]); Cox, J., Hillman, R. W., & Langevoort, D. C. (2017). Ch. 12: Inquiries into the Materiality of Information. In J. Cox, R. W. Hillman, & D. C. Langevoort (Eds.). *Securities Regulation: Cases and Materials* (8th ed.). New York: Wolters Kluwer.

21. A bright-line rule is an easily administered, straightforward rule that courts have accepted and can apply uniformly. Instead, a court's consideration of whether information is material requires a fact specific review. Vizcarra, H. (2020). The

Reasonable Investor and Climate-Related Information: Changing Expectations for Financial Disclosures. *Environmental Law Reporter, 50*(2), 10112.

22. Eccles, R. G., & Youmans, T. (2016). Materiality in Corporate Governance: The Statement of Significant Audiences and Materiality. *Journal of Applied Corporate Finance, 28*(2), 42.

23. *Basic Inc. v. Levinson*, 485 U.S. 224, 231 (1988) ("[A] minimal standard might bring an overabundance of information within its reach, and lead management simply to bury the shareholders in an avalanche of trivial information—a result that is hardly conducive to informed decisionmaking").

24. Vizcarra, H. (2019). Climate-Related Disclosure and Litigation Risk in the Oil & Gas Industry: Will State Attorneys General Investigations Impede the Drive for More Expansive Disclosures? *Vermont Law Review, 43*(4), 752–753.

25. Vizcarra, H. (2019). Climate-Related Disclosure and Litigation Risk in the Oil & Gas Industry: Will State Attorneys General Investigations Impede the Drive for More Expansive Disclosures? *Vermont Law Review, 43*(4), 733–775 (citing FindWhat Investor Group v. FindWhat.com, 658 F.3d 1282, 1305 [11th Cir. 2011]).

26. Eccles, R. G., & Youmans, T. (2016). Materiality in Corporate Governance: The Statement of Significant Audiences and materiality. *Journal of Applied Corporate Finance, 28*(2), 42.

27. United States v. Litvak, 889 F.3d 56, 64 (2d Cir. 2018) ("The standard may vary … with the nature of the traders involved in the particular market").

28. Business and Financial Disclosure Required by Regulation S-K, Concept Release, 81 Fed. Reg. 23916, 23971 (April 22, 2016) ("The role of sustainability and public policy information in investors' voting and investment decisions may be evolving as some investors are increasingly engaging on certain ESG matters …").

29. Vizcarra, H. (2019). Climate-Related Disclosure and Litigation Risk in the Oil & Gas Industry: Will State Attorneys General Investigations Impede the Drive for More Expansive Disclosures? *Vermont Law Review, 43*(4), 752.

30. Vizcarra, H. (2019). Climate-Related Disclosure and Litigation Risk in the Oil & Gas Industry: Will State Attorneys General Investigations Impede the Drive for More Expansive Disclosures? *Vermont Law Review, 43*(4), 754. at 754 (citing Meyer v. Jinkosolar Holdings Co., Ltd., 761 F.3d 245, 252 [2d Cir. 2014] [holding "ongoing and substantial pollution problems … was of substantial importance to investors" as "a reasonable investor could conclude that a substantial non-compliance would constitute a substantial threat to earnings"]).

31. Ramirez v. Exxon Mobil Corp., 334 F. Supp. 3d 832 (N.D. Tex. 2018) (plaintiffs sufficiently alleged material misstatements and loss causation in securities fraud case, allowing the case to partly survive a motion to dismiss). Other cases brought have not yet resulted in decisions. See, e.g. Complaint, Saratoga Advantage Trust Energy & Basic Materials Portfolio v. Woods, No. 3:19-cv-16380 (D.N.J. August 6, 2019), ECF No. 1.

32. People by James v. Exxon Mobil Corp., 65 Misc. 3d 1233(A) (N.Y. Sup. Ct. 2019) (slip copy). For more discussion of this case, see Vizcarra, H. (2019, December 12). Understanding the New York v. Exxon Decision. *Harvard Environmental & Energy Law Program*. Retrieved from https://eelp.law.harvard.edu/ 2019/12/understanding-the-new-york-v-exxon-decision/.

33. Vizcarra, H. (2019, December 12). Understanding the New York v. Exxon Decision. *Harvard Environmental & Energy Law Program* ("[n]o reasonable investor during the period from 2013 to 2016 would make investment decisions based on speculative assumptions of costs that may be incurred 20+ or 30+ years in the future with respect to unidentified future projects").

34. Vizcarra, H. (2019, December 12). Understanding the New York v. Exxon Decision. *Harvard Environmental & Energy Law Program*.

13

Can Investors Rely on Corporate Sustainability Commitments?

Diane Strauss and Aisha I. Saad

Abstract Companies in the United States and in Europe have been disclosing sustainability performance data in voluntary reports and in formal securities filings for decades. These reports include quantifiable ESG metrics, such as annual greenhouse gas emissions and employee diversity data, as well as aspirational targets and future commitments. In both the European and U.S. legal contexts, companies have generally been considered exempt from liability for boilerplate language and aspirational statements. Companies take advantage of this legal standard by making vague statements, limiting any commitment to concrete actions, and omitting ESG disclosures that might reflect poorly on the company. This chapter describes nascent legal frameworks in the United States and Europe that aim to require more complete ESG reporting and to protect investors against inaccurate or misleading corporate sustainability disclosures.

Keywords Sustainable investing · ESG criteria · Auditing frameworks · Sustainability reporting · Materiality · Reasonable investor · Sustainability disclosures · Materiality analysis · Credibility score

D. Strauss (✉)
Director France of Transport & Environment, Paris, France
e-mail: diane.strauss@yale.edu

A. I. Saad
Program on Corporate Governance, Harvard Law School, Cambridge, MA, USA
e-mail: asaad@law.harvard.edu

© The Author(s) 2020
D. C. Esty and T. Cort (eds.), *Values at Work*,
https://doi.org/10.1007/978-3-030-55613-6_13

Over the past half century, sustainable investing has gone from a niche practice to mainstream portfolio strategy.[1] This evolution can be seen in the growing demand for socially responsible investment funds, particularly among millennials,[2] and in the increased integration of environmental, social, and governance (ESG) considerations into mainstream investment practices.[3] Accordingly, ESG data has become more relevant to investors' strategic decision-making and portfolio governance,[4] and investors increasingly want companies to disclose decision-useful ESG data.[5]

Companies in the United States and in Europe have been disclosing sustainability performance data in voluntary reports and in formal securities filings for decades, although the form and style of corporate sustainability communications varies widely. Sustainability-related statements include quantifiable ESG metrics, such as annual carbon dioxide emissions and employee diversity data, as well as aspirational objectives and commitments. Aspirational statements are typically introduced using language such as "we hope to," "we intend to," and "we commit to," and vary widely in terms of the specificity of content and the degree of expressed commitment. In this chapter, we argue that the subjective nature of corporate sustainability statements means that they should not be treated the same by investors and by the courts. Some sustainability statements are more reliable for investors. For these statements, disclosing companies should be held liable in the case of inaccuracy or breach. Other statements are obvious window-dressing and should be dismissed as such.

In this chapter, we describe nascent legal frameworks in the United States and Europe that aim to protect investors against inaccurate or misleading corporate ESG disclosures. In both European and U.S. legal contexts, companies are generally exempt from liability for *boilerplate*[6] language and future-oriented aspirational statements. This treatment extends to published sustainability Web sites, reports, and reviews, even when investors rely on these statements in their portfolio decision-making and capital allocation. Companies take advantage of this precedent by making vague statements, eliminating commitment to concrete actions, and omitting ESG disclosures that might be useful to investors but reflect poorly on the company. Such practices flood the informational marketplace with noise, and can even overtly mislead investors and disrupt the intended functions of securities disclosure rules.

Based on a review of contemporary trends in investor decision-making, we argue that regulators, courts, and ESG data providers should take further steps to improve the reliability of corporate sustainability statements. We propose that regulators require companies to disclose the steps taken by companies to achieve their public ESG aspirations and commitments, in a narrative format similar to Management Discussion and Analysis (MD&A) reporting in form 10-K of U.S. securities filings.[7] In addition, we recommend that courts should evaluate a statement's legal actionability by testing the difference between a company's expressed commitment and the actual steps it has taken to further that commitment. Finally, ESG data providers and NGOs should play a watchdog role in this process by investigating the accuracy of company reporting and assessing the relationship between communications and actual practice.

Investors Increasingly Rely on Corporate ESG Statements

Recent surveys demonstrate that investors are increasingly relying on corporate ESG data when making portfolio decisions. A 2018 survey of mainstream investors comprising 43% of global institutional assets under management[8] found that 82% of investors consider ESG data when making investment decisions.[9] Survey results demonstrated that investors primarily use ESG information for financial motives—63% of surveyed investors reported that they consider such information to be financially material to investment performance.[10]

The rising value of assets under management dedicated to sustainable investing confirm that assessment of ESG disclosure as part of investment strategies is gaining mainstream acceptance. The Global Sustainable Investment Alliance, which tracks the annual growth of assets under management incorporating ESG strategies worldwide, notes that in the five major markets, sustainable investing comprised $30.7 trillion in 2018, a 26% increase from 2016.[11] Globally, investors with nearly $90 trillion of capital have signed on to the UN Principles for Responsible Investment.[12] Signatories of the UN Principles for Responsible Investment commit to incorporating ESG criteria into their investment decisions and to using their voting power in favor of improved corporate ESG strategies.[13]

Despite the popular demand for sustainability data, investors are still forced to rely on sustainability claims, aspirations, policies, and declared objectives that in most cases have not been verified by an independent third party. Auditing and consulting firms that produce external assurance statements for sustainability reports focus on verifying past quantitative performance metrics, such as carbon dioxide emissions and water consumption.[14] Although not widely used, more thorough auditing frameworks do exist, including the AccountAbility 1000AS sustainability reporting assurance standard, providing an example of a principles-based framework for auditing ESG information.[15] In the absence of consistent and widespread auditing, the publication of unverified sustainability reports makes it difficult for investors to distinguish a company's aspirations and intentions from concrete actions it has taken.

The gap between sustainability reporting and actual corporate behavior is exemplified by sustainability reports discussing palm oil sourcing and deforestation, a long-standing sustainability issue. In 2010, a number of large companies made public commitments to source their palm oil only from sustainable providers by the year 2020. In 2016, Greenpeace evaluated the progress made by fourteen companies and found that three of them were "on track," eight were "getting there," and three had formulated "failed promises." Colgate Palmolive, in particular, was flagged for failing to make any progress toward fulfilling its commitment. Meanwhile, the 2018 Colgate Palmolive sustainability report portrayed a different picture. The report stated, "[w]e are committed to sourcing responsible palm oil, palm kernel oil (PKO), and palm oil derivatives that do not contribute to deforestation," and "[w]e strive to meet [the objective of] full traceability of the supply chains by the end of 2020."[16] For an outside observer without access to company records, such misleading reports make it impossible to assess actual efforts. Even a conscientious investor would be hard-pressed to assess the context and validity of these statements.

The Colgate Palmolive example is not an outlier. Most reporting frameworks—including those advanced by the Sustainability Accounting Standard Board,[17] the Task Force on Climate-related Financial Disclosures,[18] and the Non-Binding European Guidelines[19]—encourage companies to formulate policies and forward-looking commitments as part of their sustainability reporting. The increase in sustainability disclosures that describe a company's policies and targets comes hand in hand with a proliferation of unreliable information, which makes it difficult for investors to distinguish between those companies that actually implement their strategies and those that do not.

Creating Legal Accountability for ESG Disclosures

The United States and Europe have developed different mechanisms for maintaining a basic level of accountability for sustainability disclosures. In the United States, securities fraud litigation allows investors to hold companies accountable for materially false or misleading ESG disclosures. In the

EU, regulation overtly requires that companies disclose specific ESG information. In both contexts, the reliability of ESG disclosures falls short of investor
needs.

U.S. Court Treatment of ESG Disclosures

The U. S. Securities and Exchange Commission (SEC), the entity responsible for regulating securities disclosures, does not mandate specific ESG
reporting.[20] Investors can contest the accuracy and completeness of sustainability statements through securities fraud litigation by claiming that certain
corporate statements were false or misleading, or that material statements
were omitted from corporate disclosures.[21] To make a successful claim, a
plaintiff must demonstrate that a statement in question was "material,"[22]
that they in fact relied on the statement, and that this reliance caused
economic loss.[23] *Materiality* is a term of art, referring to statements that
should induce reliance by a *reasonable investor*, also a term of art. Thus,
disputes over whether or not a statement is material require assumptions
about the reasonable investor and their informational needs and investment
behavior.[24]

While securities fraud case law comprises a robust body of legal doctrine,
cases that specifically focus on sustainability disclosures remain limited.
In at least two examples, however, federal courts have determined that
sustainability statements were material to the reasonable investor. In a 2012
securities case against BP, plaintiffs alleged that the company's statements
concerning the magnitude of its Deepwater Horizon oil spill, and its reaction to the disaster, were misleading.[25] The court examined sustainability
statements in BP corporate securities filings, public statements, and other
less formal disclosures. It found that repeated statements from BP executives concerning the implementation of safety improvements were indeed
misleading and actionable. In another securities case, against *Massey Energy
Co.*,[26] the court found that statements concerning employees' occupational
safety were material and misleading. In both of these cases, the statements in
question were repeated several times, were concrete enough to be measured
against a clear benchmark, and met the standard of being misleading to a
reasonable investor.

In a recent article, we identified three key criteria that courts consider
when determining whether or not a corporate statement is material: (1) the
type of document where the statement appears, (2) the statement's generality or specificity, and (3) the statement's time posture.[27] First, we noted
that courts are inclined to consider statements made in a variety of formats

and publications as potentially material, provided they informed a reasonable investor. Judges have found that content in press releases, investment prospectuses, news articles, publicly filed annual reports,[28] financial statements, audit reports,[29] Web sites,[30] and technical advertisements[31] may be legally actionable. When focusing specifically on the actionability of sustainability statements, however, the case law remains thin. The BP case demonstrates that sustainability statements can be found actionable when published outside of financial reports but does not make a categorical determination that all such statements are actionable. In the case of *Bondali vs Yum!*,[32] by contrast, the court found that statements made in company codes of conduct were not actionable because such publications are inherently aspirational. But many company codes of conduct should not be taken as a perfect analogy for sustainability reports. While both contain certain aspirational elements, sustainability reports also include information concerning actual company operations or targets. Thus, sustainability reports and reviews should not be dismissed as categorically aspirational.

Second, we observed that courts tend to examine generality and specificity of language when determining a statement's actionability. An article by Ajax and Strauss concluded that a statement's form was the most important criterion considered in a court's review of actionability.[33] Statements that are concrete and measurable are likely to be found actionable by the courts, while vague, aspirational, and optimistic claims are dismissed as a type of inactionable speech termed "puffery." The puffery defense provides legal cover for statements that are too obviously general for any reasonable investor to rely on them.[34] Such statements might include language like "we strive to" or "we intend to," or hyperboles describing a company's features or operations. The majority of U.S. circuits have identified generality or vagueness as criteria for distinguishing puffery from actionable statements[35] and have determined that puffery is inactionable as a matter of law.

Third, we found that courts tend to emphasize a statement's time posture when making a judgment on its materiality. While courts tend to find that statements about the past and present can be potentially misleading, they systematically treat forward-looking statements as an expression of optimism that is unlikely to inform the decisions of reasonable investors. Defendants rely on the 1995 Private Securities Litigation Reform Act[36] to prevent legal action for forward-looking statements. The Private Securities Litigation Reform Act provides legal cover to companies by granting a *safe harbor* to forward-looking statements. To receive safe harbor treatment, statements must be accompanied by meaningful, cautionary language identifying factors which could cause outcomes to differ materially from the forward-looking claims. The 2018 case of *Ramirez v. Exxon Mobil Corp.* provides some insight into the court's application of the Private Securities Litigation Reform Act in the context of forward-looking sustainability statements. In *Ramirez*, New York's Attorney General accused ExxonMobil of misleading investors by publicly disclosing an estimated *proxy cost* of greenhouse gases. At the same time, Exxon was using lower numbers internally, thereby miscon-struing the public evaluation of the company's reserves. The carbon proxy comprised a quantifiable statement that had direct implications for the firm's balance sheet. Importantly, the court examined actual investor reliance on the forward-looking statement rather than dismissing the allegations outright as a matter of law.[37] This decision suggests that the Private Securities Litigation Reform Act does not necessarily immunize forward-looking ESG statements from liability.

At present, securities fraud litigation on its own remains an inadequate tool to ensure reliable ESG disclosures. This could be remedied, at least in part, by regulations from the SEC requiring standardized and system-atic disclosure of ESG data. Such regulation could also include a narrative reporting requirement, similar to Management Discussion and Analysis (MD&A) narrative reporting in form 10-K filings. In this narrative section, companies might provide a contextualized account of concrete steps taken toward their expressed sustainability aspirations and commitments, thereby providing investors a more discernible account.

Trends in European Treatment of ESG Statements

The European Union, in contrast to the United States, has favored a regula-tory path to creating accountability for ESG statements. In 2014, the Euro-pean Union enacted the Non-Financial Reporting Directive, requiring that companies with more than 500 employees disclose a non-financial statement along with their financial filings.[38] This law applies to nearly 6000 compa-nies. The non-financial statement should include "at least environmental matters, social and employee-related matters, respect for human rights, anti-corruption and bribery matters."[39] The Directive explicitly requires disclo-sure of a selection of environmental and social data, including the company's share of renewable energy, water and air pollution, gender equality, and prevention of human rights abuses.[40] It also requires that companies describe principal risks and policies associated with sustainability, risks of adverse impacts, and due diligence processes.[41]

In drafting the Directive, European policymakers sought to incorporate some flexibility in the scope and quality of corporate reporting. This flexibility is reflected in a number of key provisions. First, the Directive allows European companies to avoid complying with their disclosure obligation if they provide a reason for doing so under a principle of "comply and explain." Second, the Directive explicitly allows companies to choose from among dozens of reporting frameworks, including the Global Reporting Initiative, the UN Global Compact, the Organisation for Economic Co-operation and Development (OECD)[42] Guidelines for Multinational Enterprises, and the ISO 26000 management system standard for sustainability. Third, the Directive is enforced by member states rather than by the European Union. While these provisions provide helpful flexibility for companies, they also result in widely divergent reporting practices which make it difficult for investors to compare ESG data between companies.[43] Furthermore, the task of ensuring that issuers comply with the Directive falls on individual member states and their enforcement agencies, and their willingness to sanction disclosure violations.

In a 2018 assessment of the Directive's first year of implementation, the European Securities and Markets Authority reviewed non-financial statements from 386 companies along with national enforcement measures.[44] The assessment found that all issuers had complied with their obligations to provide some sort of statement. However, roughly a quarter of the disclosure items were deemed unsatisfactory, with 27% of issuers providing boilerplate language and 25% of issuers failing to provide at least one piece of mandatory information. The assessment also counted 51 enforcement measures relating to non-financial statements across Europe. In France, for example, the Autorité des Marchés Financiers, the enforcement agency regulating French financial markets, is responsible for ensuring that management companies produce quality information about their strategies for combating climate change.[45] A recent provision, the "duty to due diligence,"[46] permits consumers, NGOs, employees, and suppliers to take legal action against companies for misstatements regarding environmental and social impacts. In light of this provision, European Member States may play a role in ensuring greater corporate accountability for their ESG statements, both through regulatory pressure and by providing recourse to investors and consumers through securities law and consumer law.

As illustrated above, the United States and the European Union have pursued different strategies to sanction sustainability disclosures. In the United States, courts play an important role in ensuring the reliability of concrete, measurable facts that describe corporate policies and performance, but current case law precedent allows companies to evade liability under a puffery cover for vague, forward-looking commitments and aspirational statements. The European Union, on the other hand, focuses on the completeness of information by listing specific issues and processes that must be disclosed. While boilerplate language is considered unsatisfactory disclosure by the European Securities and Markets Authority, such statements are not explicitly banned by law. The success of the EU approach to data reliability largely depends on the implementing role of domestic enforcement. In both the United States and the European Union contexts, the current legal framework remains too weak to ensure consistent and reliable ESG information for investors.

Toward Greater Accountability for Corporate ESG Disclosures

Contemporary investors' need for reliable ESG data calls for regulatory reform of corporate disclosure. Such reform should improve the credibility of corporate ESG statements, while still allowing corporate management the flexibility to adjust their sustainability strategies over time. We propose that regulators, courts, and financial markets limit the discrepancy between the language in a corporate statement or disclosure, and the actionable steps toward its implementation. This may be achieved when companies provide a thorough account of their present policies and actions taken to fulfill their ESG commitments. The Securities and Exchange Commission and the European Commission should require that companies report annually on their policies and commitments, either updating investors on steps taken in furtherance of these commitments or explicitly abandoning them. For investors, this approach would provide the information needed for more accurate assessment of company performance and risk. For companies, this would provide liability protection against claims based on investor reliance beyond the expectations outlined in the company's reports. This requirement would strengthen investors' trust in ESG commitments without binding corporate management to their targets.

In the absence of mandatory disclosure requirements, courts in the United States and in Europe might protect investors against corporate fraud and misstatements by adopting a balancing test that focuses on the discrepancy between the expectations induced by an ESG statement and the substance of its underlying facts. This test would focus on the commonalities and differences between a public commitment and the associated corporate practice. The materiality analysis would then balance a statement's generality and tone, and the severity and pervasiveness of an underlying corporate violation.[47] This approach would draw new lines in the legal determination of materiality and reliance that are better aligned with contemporary investor expectations of ESG disclosures. Companies would be incentivized to systematically disclose their efforts and progress toward commitments to prevent legal action, and would refrain from formulating empty or misleading statements.

Finally, third-party entities, such as ESG ratings firms, credit agencies, and accounting firms, can take a greater role in screening corporate commitments. Thus far, NGOs have played a key role in strategically reviewing the implementation of some corporate ESG commitments. They have focused on critical social and environmental issues, but their role is not to systematically review all corporate ESG commitments. Third-party entities might expand this reviewing role given investor demand for such content. While ESG ratings firms incorporate forward-looking commitments into their scores, we the authors are not aware of examples in which they systematically track the implementation of policies pursued to achieve these commitments. A "credibility score" may emerge from this review, where companies using boilerplate language and failing to translate commitments into actions would score lower than companies acting on their commitments and regularly updating investors on their progress. This credibility score would evaluate the ambition of sustainability commitments, progress as indicated by correspondence between statements and action, and transparency in reporting.

In the absence of clear, mandatory, strictly enforced ESG reporting requirements, U.S. and European investors currently have to gauge the credibility of corporate commitments themselves. Investor appetite for ESG data supports the need for a coherent ESG reporting framework that allows for greater consistency and reliability. This framework may be brought about through regulatory reforms that improve the availability and consistency of ESG data, legal reforms that play a gatekeeping role in controlling the quality and accuracy of ESG disclosures, or third-party engagement that provides auditing and credibility.

Notes

1. A 1975 decision by the Securities and Exchange Commission dismissed social disclosure campaigns as irrelevant, elaborating that ethical funds comprised a negligible 0.0005% of mutual fund assets in the United States.

By 2018, U.S. domiciled assets under management incorporating sustainability strategies comprised $12 trillion, representing 26% of total assets under professional management. *Nat. Res. Def. Council, Inc. v. Sec. & Exch. Comm'n*, 606 F.2d 1031, 1039 (D.C. Cir. 1979) cited and discussed in Williams, C. A. (1999). The Securities and Exchange Commission and Corporate Social Transparency. *Harvard Law Review, 112*(6), 1197–1311. Retrieved from https://www.jstor.org/stable/pdf/1342384.pdf; US SIF Foundation. (2018). *Report on US Sustainable, Responsible and Impact Investing Trends 2018*. Retrieved from https://www.ussif.org/files/Trends/Trends%202018%20executive%20summary%20FINAL.pdf. Globally, a 2016 report by the GSI Alliance found that SRI comprised 26.3% of Total Managed Assets. Global Sustainable Investment Alliance. (2016). *2016 Global Sustainable Investment Review*. Retrieved from http://www.gsi-alliance.org/wp-content/uploads/2017/03/GSIR_Review2016.F.pdf..

2. Millennial investors are twice as likely as other investors to invest in companies incorporating ESG practices. Ruggie, J. G., & Middleton, E. K. (2018). *Money, Millennials and Human Rights—Sustaining "Sustainable investing"*. Harvard Kennedy School Mossavar-Rahmani Center for Business and Government. Retrieved from https://www.hks.harvard.edu/sites/default/files/centers/mrcbg/working.papers/CRI69_FINAL.pdf.

3. One notable example is Laurence Fink's annual letter to companies. Fink is the Chairman and CEO of BlackRock, the world's largest investment firm. Fink's annual letter represents a proxy for the pulse of the mainstream investment world. In 2018, 2019, and 2020, his letter has emphasized the importance of ESG and sustainability in investing. Fink, L. (2018). A Sense of Purpose. *BlackRock*. Retrieved from https://www.blackrock.com/corporate/investor-relations/2018-larry-fink-ceo-letter; Fink, L. (2019). Profit & Purpose. *BlackRock*. Retrieved from https://www.blackrock.com/americas-offshore/2019-larry-fink-ceo-letter; Fink, L. (2020). A Fundamental Reshaping of Finance. *BlackRock*. Retrieved from https://www.blackrock.com/corporate/investor-relations/larry-fink-ceo-letter.

4. ISS analytics documented over 140 environmental and social proposals centered on disclosure, risk assessment, and oversight in 2018, up from 60 such proposals in 2000. Mishra, S. (2018, February 28). An Overview of U.S. Shareholder Proposal Filings. *Harvard Law School Forum on Corporate Governance*. Retrieved from https://corpgov.law.harvard.edu/2018/02/28/an-overview-of-u-s-shareholder-proposal-filings/; Global Sustainable Investment Alliance. (2018). *2018 Global Sustainable Investment Review*. Retrieved from http://www.gsi-alliance.org/wp-content/uploads/2019/03/GSIR_Review2018.3.28.pdf.

5. As of September 2019, the Task Force on Climate-related Financial Disclosures was supported by 876 organizations, most of them financial institutions. Task Force on Climate-related Financial Disclosures. (2020). TCFD Supporters. *TCFD*. Retrieved from https://www.fsb-tcfd.org/tcfd-supporters/.

6. In this context, the term boilerplate refers to text in a disclosure that is standardized or generic. Boilerplate language is commonly used for efficiency and to increase standardization in the structure and language of legal documents, such as contracts, investment prospectuses, and bond indentures.

7. U.S. Securities and Exchange Commission. (2008). Financial Reporting Manual. *U.S. Securities and Exchange Commission.* Retrieved from https://www.sec.gov/corpfin/cf-manual/topic-9 ("MD&A is a narrative explanation of the financial statements and other statistical data that the registrant believes will enhance a reader's understanding of its financial condition, changes in financial condition and results of operation.").

8. Amel-Zadeh, A., & Serafeim, G. (2018). Why and How Investors Use ESG Information: Evidence from a Global Survey. *Financial Analysts Journal, 74*(3), 87–103. Retrieved from https://www.cfainstitute.org/en/research/financial-analysts-journal/2018/faj-v74-n3-2.

9. Amel-Zadeh, A., & Serafeim, G. (2018). Why and How Investors Use ESG Information: Evidence from a Global Survey. *Financial Analysts Journal, 74*(3), 88.

10. Amel-Zadeh, A., & Serafeim, G. (2018). Why and How Investors Use ESG Information: Evidence from a Global Survey. *Financial Analysts Journal, 74*(3), 91.

11. Global Sustainable Investment Alliance. (2018). *2018 Global Sustainable Investment Review.* Retrieved from http://www.gsi-alliance.org/wp-content/uploads/2019/03/GSIR_Review2018.3.28.pdf.

12. Principles for Responsible Investment. (2020). *PRI.* Retrieved from www.unpri.org.

13. While these numbers are widely quoted, other evidence suggests that even though policy commitments are quite large, including under the PRI, actual ESG assets under management fall short of those commitments. Hale, J. (2019). *Sustainable Funds U.S. Landscape Report: More Funds, More Flows, and Strong Performance in 2018.* Morningstar. Retrieved from https://s3.amazonaws.com/v3-app_crowdc/assets/4/4e/4e4caadd3df08d46/2018_Sustainable_Funds_U.S._Landscape_Report.original.1557166295.pdf?1557166298.   The general trend remains one of upward growth in sustainable investment, but this is a gap worth noting.

14. Sridharan, V., & Saush, A. (2020). *Sustainability Matters: Bridging the ESG Disclosure Gap.* The Conference Board. Retrieved from https://www.conference-board.org/pdfdownload.cfm?masterProductID=20672.

15. AccountAbility, AA1000 AccountAbility Principles (AA1000AP) 2018. https://www.accountability.org/standards/.

16. Colgate-Palmolive Co. (2020). Our Policy on Responsible and Sustainable Sourcing of Palm Oils. *Colgate-Palmolive.* Retrieved from https://www.colgatepalmolive.com/en-us/core-values/our-policies/palm-oils-policy.

17. Sustainability Accounting Standards Board. (2017). *SASB Conceptual Framework.* Retrieved from https://www.sasb.org/wp-content/uploads/2019/

05/SASB-Conceptual-Framework.pdf. "[T]he SASB's approach to sustainability accounting consists of defining operational metrics on material, industry-specific sustainability topics likely to affect current or future financial value. Like financial accounting information, sustainability accounting information captures past and current performance, and can also be forward-looking to the extent that it helps management describe known trends, events, and uncertainties that may reveal an actual or potential impact on the financial condition or operating performance of a reporting entity."

18. Task Force on Climate-related Financial Disclosures. (2020). Recommendations Overview. *TCFD*. Retrieved from https://www.tcfdhub.org/recommend ations/. "The TCFD recommendations are designed to solicit consistent, decision-useful, forward-looking information on the material financial impacts of climate-related risks and opportunities, including those related to the global transition to a lower-carbon economy."

19. European Commission. (2017). Communication from the Commission: Guidelines on non-financial reporting (methodology for reporting non-financial information). *Official Journal of the European Union*. Retrieved from https://ec.europa.eu/anti-trafficking/sites/antitrafficking/files/guidel ines_on_non-financial_reporting.pdf. "Forward-looking information enables users of information to better assess the resilience and sustainability of a company's development, position, performance and impact over time. It also helps users measure the company's progress towards achieving long-term objectives."

20. The closest that the S.E.C. has come to mandatory ESG disclosure is its 2010 Guidance Regarding Disclosure Related to Climate Change. This guidance was advisory and non-binding. U.S. Securities and Exchange Commission. (2010, February 8). *Commission Guidance Regarding Disclosure Related to Climate Change, Release No. 33-9106*. Retrieved from https://www.sec.gov/rules/interp/2010/33-9106.pdf.

21. Under *Basic Inc. v. Levinson*, the Supreme Court adopted the standard for materiality set out in the case ot *TSC Indus., Inc. v. Northway, Inc.*, 426 U.S. 438 (1976) to Section 10-b disclosures. *Basic Inc. v. Levinson*, 485 U.S. 224, 231–32 (1988) ("[T]o fulfill the materiality requirement 'there must be a substantial likelihood that the disclosure of the omitted fact would have been viewed by the reasonable investor as having significantly altered the "total mix' of information made available."). In other words, companies who choose to disclose can't omit part of the information if this omission makes the statement misleading.

22. *TSC Indus., Inc. v. Northway, Inc.*, 426 U.S. 438, 445 (1976) ("The question of materiality, it is universally agreed, is an objective one, involving the significance of an omitted or misrepresented fact to a reasonable investor. Variations in the formulation of a general test of materiality occur in the articulation of just how significant a fact must be or, put another way, how certain it must be that the fact would affect a reasonable investor's judgment.").

23. In order to make a successful section 10(b) claim, a plaintiff must prove "(1) a material misrepresentation or omission by the defendant; (2) scienter; (3) a connection between the misrepresentation or omission and the purchase or sale of a security; (4) reliance upon the misrepresentation or omission; (5) economic loss; and (6) loss causation." *Stoneridge Inv. Partners, LLC v. Sci.-Atlanta*, 552 U.S. 148, 157 (2008).

24. *TSC Industries, Inc. v. Northway, Inc.* enshrined the "reasonable investor" in securities doctrine as the litmus test for determining whether corporate disclosures and other statements are material or immaterial for the purposes of section 10(b) litigation. Accordingly, the notions of materiality and the reasonable investor go hand-in-hand. 426 U.S. 438 (1976).

25. *In re BP p.l.c. Sec. Litig.*, 843 F. Supp. 2d 712 (S.D. Tex. 2012).

26. *In re Massey Energy Co. Sec. Litig.*, 883 F. Supp. 2d 597, 613 (S.D.W. Va. 2012).

27. *McGann v. Ernst & Young*, 102 F.3d 390, 397 (9th Cir. 1996); *S.E.C. v. Rana Research, Inc.*, 8 F.3d 1358, 1362 (9th Cir. 1993).

28. *In re Ames Dep't Stores Inc. Stock Litig.*, 991 F.2d 953, 968 (2d Cir. 1993).

29. *Semerenko v. Cendant Corp.*, 223 F.3d 165 (3d Cir. 2000).

30. *In re Plains All Am. Pipeline, L.P. Sec. Litig.*, 245 F. Supp. 3d 870, 900 (S.D. Tex. 2017) ("[T]he third statement—that Plains makes repairs and replacements when needed on *all* of its pipelines—is actionably misleading…. This website statement is actionable and material.").

31. *In re Carter-Wallace, Inc. Sec. Litig.*, 150 F.3d 153, 156 (2d Cir. 1998) ("Technical advertisements in sophisticated medical journals detailing the attributes of a new drug could be highly relevant to analysts evaluating the stock of the company marketing the drug.").

32. *Bondali v Yum! Brands Inc.*, 2015 WL 4940374 (6th Cir. Aug. 20, 2015)— 1.05[2], 1.05[5], 6A.04[1].

33. Ajax, C. M., & Strauss, D. (2019). Corporate Sustainability Disclosures in American Case Law: Purposeful or mere "Puffery"? *Ecology Law Quarterly*, 45(4), 703–734.

34. *Phoenix Payment Solutions, Inc. v. Towner*, 2009 U.S. Dist. LEXIS 91978 (D. Ariz. Oct. 2, 2009) (Puffery refers to statements that are "so exaggerated, vague, or loosely optimistic" that they are "immaterial and unworthy of reliance" and "cannot serve as the basis for liability".)

35. *Shaw v. Digital Equip. Corp.*, 82 F.3d 1194, 1217 (1st Cir.1996); *Lasker v. New York State Elec. & Gas Corp.*, 85 F.3d 55, 59 (2d Cir. 1996); *Raab v. General Physics Corp*, 4 F.3d 286, 289–290 (4th Cir. 1993); *Nathenson v. Zonagen, Inc.* 267 F.3d 400, 404, 419 (5th Cir.2001); *In re Ford Motor Co. Sec. Litig., Class Action*, 381 F.3d 563, 570–571 (6th Cir. 2004); *Searls v. Glasser*, 64 F.3d 1061,1066 (7th Cir. 1995); *In re K–tel Int'l, Inc. Sec. Litig.*, 300 F.3d 881, 897 (8th Cir.2002); *Grossman v. Novell, Inc.* 120 F.3d 1112,1119 (10th Cir.1997).

36. Refers to the Private Securities Litigation Reform Act which was intended to prevent frivolous securities lawsuits. 15 U.S.C.A. § 78u-5 (West). The

PSLRA distinguishes between statements that focus on the past and present, and those that are forward-looking. It grants a "safe harbor" to forward-looking statements, making them inactionable.

37. *Ramirez v. Exxon Mobil Corp.*, 334 F. Supp. 3d 832, 850 (N.D. Tex. 2018) ("Courts cannot apply a blanket safe harbor for all forward-looking statements but must determine how a statement is specifically and meaningfully protected by the safe harbor.... Boilerplate cautionary language does not provide a substantive and meaningful warning."). Ultimately, however, the claim was dismissed as failing to meet the standard of significantly altering the "total mix" of information available to a reasonable investor.

38. European Commission. (2020). Non-Financial Reporting. *European Commission*. Retrieved from https://ec.europa.eu/info/business-economy-euro/com pany-reporting-and-auditing/company-reporting/non-financial-reporting_en.

39. National legislative chambers may decide to reinforce the provisions of the Directive but they are not allowed to weaken these provisions.

40. Recital No7 of Directive 2014/95/UE of 22 October 2014. "Where undertakings are required to prepare a non-financial statement, that statement should contain, the environmental, and, as appropriate, on health and safety, the use of renewable and/or non-renewable energy, greenhouse gas emissions, water use and air pollution; (…) gender equality, implementation of fundamental conventions of the International Labor Organization, working conditions, social dialogue, respect for the rights of workers to be informed and consulted, respect for trade union rights, health and safety at work and the dialogue with local communities, and/or the actions taken to ensure the protection and the development of those communities (…) [as well as] information on the prevention of human rights abuses and/or on instruments in place to fight corruption and bribery." https://eur-lex.europa.eu/legal-content/EN/TXT/?uri=CELEX% 3A32014L0095.

41. Official Journal of the European Union, Directive 2014/95/EU of the European Parliament and of the Council of 22 October 2014. https://eur-lex.europa.eu/ legal-content/EN/TXT/PDF/?uri=CELEX:32014L0095&from=EN.

42. OECD: Organisation for Economic Co-operation and Development.

43. The European Securities Market Agency (ESMA) "recommends revisiting the significant optionality that currently characterizes the NFRD to provide for a more stringent set of requirements, thus promoting consistency in disclosure and enforcement practices." https://www.esma.europa.eu/sites/default/files/lib rary/esma32-334-109_comment_letter_on_revision_of_ec_nbg_on_non-fin ancial_reporting.pdf.

44. European Securities and Markets Authority. (2019). *Final Report: ESMA's Technical Advice to the European Commission on Integrating Sustainability Risks and Factors in MiFID II*. Retrieved from https://www.esma.europa.eu/sites/def ault/files/library/esma35-43-1737_final_report_on_integrating_sustainability_ risks_and_factors_in_the_mifid_ii.pdf.

45. Autorité des Marchés Financiers. (2020). The AMF in Figures. *AMF.* Retrieved from https://www.amf-france.org/en/amf/presentation-amf/amf-figures.
46. In French, the devoir de vigilance, European Coalition of Corporate Justice. (2017). French Corporate Duty of Vigilance Law (English translation). *Respect.* Retrieved from http://www.respect.international/french-corporate-duty-of-vigilance-law-english-translation/.
47. Lipton, A. M. (2017). Reviving Reliance. *Fordham Law Review, 86* (1), 91–147.

14

Financial Regulations and ESG Investing: Looking Back and Forward

Michael Eckhart

Abstract Regulations and government policy are key elements in the success and growth of environmental, social, and governance (ESG) investing. *Regulation* refers to a broad array of mandatory requirements, guidelines, incentives, and other types of policy intended to steer the flow of capital, not just by government, but also by industry. This chapter explores the current state of play of relevant regulations, looking specifically at how targeted regulations and policies enabled clean energy—today's biggest sector of sustainable investing representing about $300 billion per year since 2010. The chapter concludes with an outlook on where regulations and government policies supporting, and sometimes blocking, ESG-screened investing are likely to proceed in the coming decade.

Keywords Sustainable investing · Investment capital flow · Risk and return · Investor-Owned Utilities · Blended finance · Socially responsible investing · ESG-screening · Climate change risk · Green finance · Regulated markets · Green bonds · Policy-business models · Public policy

Regulations and government policy are a key element in the success and growth of environmental, social, and governance (ESG) investing. *Regulation* refers to a broad array of mandatory requirements, guidelines, incentives, and

M. Eckhart (✉)
University of Maryland, College Park, MD, USA

© The Author(s) 2020
D. C. Esty and T. Cort (eds.), *Values at Work*,
https://doi.org/10.1007/978-3-030-55613-6_14

other types of policy intended to steer the flow of capital, not just by government but also by industry. This chapter explores the current state of play of relevant regulations, looking specifically at how targeted regulations and policies enable clean energy—today's biggest sector of ESG investing, representing about $300 billion per year since 2010. The chapter concludes with an outlook on where regulations and government policies supporting ESG investing are likely to proceed in the coming decade.

While regulations have undoubtedly influenced ESG investing, government regulators do not dictate what investors and lenders do with their money (with a few exceptions, such as Russia and China). Rather, they tend to regulate the mechanisms of investment capital flow, as well as the disclosure of information to give investors and lenders adequate information for their decisions. However, this agnostic approach to regulation has lately shifted somewhat as governments seek to meet their commitments under climate change and other social obligations.

Most financial professionals seek to manage capital in order to achieve selected outcomes on a scale of *risk and return*. Some examples of investment products along this spectrum are:

- *Low risk/low returns*: U.S. Treasury Bonds (currently yields 0.5–2.5% in the United States and negative yields in Europe).
- *Modest risks/modest returns*: Investment-grade corporate bonds (currently 2.5–5%).
- *Medium risk/medium returns*: Publicly traded equities (a.k.a. the "stock market") where the average return over many decades has been about 7%.
- *High risk/high returns*: Private equity investments seeking returns greater than 15%, and venture capital seeking rates of return of 40% and more.

These categories of risk and return are well established in the United States and Western financial markets. But balancing risk and return becomes more complex when looking at investments and lending activities around the world and especially in developing countries where risk factors are generally higher—or even unmeasurable. The time horizon of infrastructure investments becomes another point of speculation to many—especially bank regulators. In the face of these complexities, Multilateral Development Banks (MDBs) and other Development Finance Institutions (DFIs) have risen to enable investment by backing governments to mitigate risk.

The primary role of government regulation and policy is to directly or indirectly influence investors' balancing of risk and return. Investment and financing are implemented through many kinds of contracts, which are of course interpreted under the rule of laws that vary by state, province, and country. For example, many perceive that the laws of the State of New York are best for contracts and financing with over 100 years of court decisions, while the laws of the State of Delaware are deemed best for incorporation of a company with limited liability to officers and board members.[1] Regulations can establish boundaries, while other policies can provide investors with steerage within existing laws, and they can go so far as to change and create new laws. Investors seek the clarity and opportunity created by laws, regulations, policies, and practices that are established, stable, and proven. It is often not enough to enact a new law or regulation—they must stand the test of use, challenges, and time.

The U.S. wind power industry provides a good example of how policy and regulatory differences can determine outcomes. To produce electricity at a low cost over time, a wind farm needs to spread its front-end capital cost over 15–20 years. Under U.S. banking regulations, a deposit-taking bank— e.g., Bank of America, Citibank, J. P. Morgan, Wells Fargo—cannot make a business loan with a tenor (duration) longer than seven years. However, regulators in Europe, Japan, and South Korea allow their banks to lend for more than 20 years if the funds are for high-quality assets in a low-risk market such as U.S. wind farms. As a result of these regulatory differences, much of the senior debt lent to U.S. wind projects has not come from U.S. banks, but from European, Japanese, and South Korean banks.

Another example of policy-driven investment is revealed by the different ownership models around the world of electric utilities that buy renewable power. In the United States, most of the population is served by Investor-Owned Utilities (IOUs), which have several advantages for investors:

- The shares of IOUs are publicly traded, offering transparency on their financial status;
- IOUs are regulated at the state level, providing a consistent operating environment;
- IOUs have a lengthy track record (sometimes 100 years or more) on which to assess risk and return; and
- IOUs maintain credit ratings (risk of default) by Standard & Poors, Moody's, Fitch, and other rating agencies.

Investors and lenders can therefore "look through" the wind and solar projects—and how risky they may or may not be—to the credit worthiness of the "off-take" utility, which buys the electricity, to determine the risk of such investment or loan in a project.

Comparatively, government-owned electric utilities serve most populations outside of the United States, including in China and India. Almost all emerging economies generally lack the reassuring traits of U.S. utilities that allow investors to accept riskier wind and solar generation when a stable source of revenue from an established off-take utility is present. Thus, the same wind farm in the United States will have higher—and often much higher—perceived risks if proposed in a developing country.

There has been a lag in renewable energy investment in emerging markets.[2] Much of this lag is likely the result of developing countries' governments' failing to address and mitigate the financial risks of climate-friendly projects. Some argue that because of the urgency of climate change, investors and lenders should and must step up and assume more risk. But I argue that because the beneficiary of mitigating climate change is society as a whole—not the specific investor or lender—society, as represented by governments, should assume, mitigate, and eliminate such higher risks for investors to be able to profit from climate change mitigation. In parallel, there has been a rising recognition that the public sector simply does not have the financial resources to achieve the visions set out by the Paris Climate Accord and the UN Sustainable Development Goals (SDGs).[3] Governments have therefore concluded that the world must look to the private sector and the financial sector to redirect capital to the desired purposes.

Governments have two strategies to undertake this risk mitigation. The first is to simply accept (and pay for) the risks by, for example, providing a guarantee of a loan or guarantee that the off-taker utility will pay its bills on time. The second is to change policy, including enacting new regulations, to reduce or eliminate the apparent risks. This transformation might be accomplished by changing utility ownership from government-owned to investor-owned, and then putting in place the necessary regulatory apparatus and financial practices to provide a stable investment environment. The initial planning efforts concluded that we must develop methods of *blended finance* to bring public and private money together to fund ESG investments and climate change solutions.

There are hundreds of policy actions that can be taken to increase investment and financing of climate-friendly and ESG-oriented solutions. In the following sections, I summarize a variety of actions that governments have or could enact to enable and facilitate more responsible investments by the private sector.

Evolution of Regulations and Policy that Support Sustainable Investing

The movement to address social challenges through investment actually began in the 1800s when the Methodist Church crafted guidelines on the ethics and morality of investing, and resurged during the 1980s South African apartheid divestment movement. Sustainable finance was first formalized in 1982 in the form of socially responsible investing (SRI), using a method of negative screening to identify antisocial behavior by companies such as weapons production, tobacco manufacturing, and environmental pollution. Socially responsible investing started with the founding of Trillium Asset Management by Joan Bavaria in Boston, and in 1983 with the founding of Winslow Management by Jackson W. Robinson in New York.

In parallel, the modern field of sustainability is said to have started with the publishing of the book *Silent Spring* in 1962 by Rachel Carson, which scientifically documented how the pesticide DDT was killing off birdlife including the bald eagle, hitting at the heart of American culture. It picked up again in the early 1970s with the enactment of U.S. laws including the Clean Air Act, Clean Water Act, and Endangered Species Act, and the creation of the United States Environmental Protection Agency, followed by a rapid series of laws and regulations protecting nature and humans from harmful chemicals, habitat destruction, and pollution. The field continued to evolve as evidence emerged about climate change caused by human emissions of carbon dioxide from the combustion of fossil fuels and emissions of methane from oil and gas extraction, coal mining, and agriculture.

In response to the global threat of climate change, the United Nations established the Intergovernmental Panel on Climate Change in 1990 and the Framework Convention on Climate Change in 1992. These organizations have supported and promulgated climate change policy models for governments at the national and local level leading up to the Kyoto Protocol in 1996, and the subsequent Paris Climate Change Agreement in 2015. Over this entire time period, especially since the Paris Agreement, the investment community continued trying to address climate change and other ESG issues. Subsequently, an upgraded form of investment analysis termed "ESG Investing" took hold in the 2000s. The International Capital Market Association has managed the adoption of specialty guidelines such as the Green Bond Principles, originally by a group of 13 bond underwriting banks in 2014, followed by the Social Bond Principles, the Sustainable Bond Principles, and others. These guidelines represented capital market self-imposed methods and guidelines by investors and financial institutions rather than government regulations.

Governments began taking concrete action on climate change following the Paris Climate Agreement and the mandate for countries to adopt Nationally Determined Contributions (NDCs) outlining the actions they would take to help achieve the goals of limiting global warming to well below two degrees Celsius above pre-industrial levels. Under the Agreement, governments committed to report progress on achieving their NDCs at the 2020 Conference of the Parties (COP), but this conference was delayed by government restrictions on travel and meetings due to the COVID-19 pandemic that year. The NDCs, while not government regulations in the traditional sense, serve as a fundamental underlying driver of future regulations and ESG investing.

Similarly, the UN Sustainable Development Goals (SDGs) act as an informal policy driver for ESG investing. Governments seeking to address the SDGs and achieve their NDCs have begun to enact laws, regulations, and policies that drive their citizens and companies to take individual action and drive the financial sector to support those actions with investment and lending. For example, China enacted a Green Bond Law, Europe adopted an EU Green Bond Standard, and Japan adopted a Green Finance Network, among many others.

Current State of Play in Regulations and Policy that Support Sustainable Investing

Policymakers are still in the early days of learning how to regulate, influence, and guide the financial sector on sustainable investing and lending. In 2012, Larry Fink, founder and CEO of BlackRock, the world's largest money manager, was asked in an interview at a TED Talk conference how he thought about "climate risk." His response was that he "hadn't thought about that one." Eight years later, in 2020, Fink wrote in his annual letter to CEOs that "climate risk is investment risk," indicating the speed at which the financial system is changing.[4] Fink's directives to his fund managers could achieve the same outcome as government rules in moving capital away from climate change risk.

One of the key barriers to this emerging focus on climate risk are the banking regulations themselves. In 2009, international banking regulations were agreed in Basel, Switzerland (termed Basel III as they were the third such agreement negotiated in the city). The Basel III accord introduced a set of reforms designed to mitigate risk within the international banking sector, by requiring banks to maintain proper leverage ratios and keep certain levels of reserve capital on hand.[5] One of the outputs of the agreement was to preclude banks from undertaking long-term lending, as this practice was deemed high-risk in the wake of the 2008 financial crisis. Six years later, governments from around the world agreed to limit greenhouse gas emissions in Paris at the conclusion of the COP 21 conference, implicitly calling for long-term actions with long-term financing. We are in a transition and we now live in a world of Basel III versus COP 21.

The regulatory scheme affecting ESG investing is vast and complex, and so deeply rooted in historical path-dependency that any changes are difficult to make. Today, in addition to the SDGs and Paris Climate Change Accord, there exists a wide spectrum of government regulation and guidelines that impact the flow of capital from bad to good, brown to green, and carbon-emitting to climate change-mitigating. There are also a wide array of voluntary guidelines and tools available to companies and investors to guide strategy and disclosure. This complex ecosystem of regulation, policies, guidelines, standards, and frameworks is explored in the following sections.

Global Guidelines

A number of guidelines have emerged to provide learning and tools to investors interested in pursuing ESG or sustainable investing strategies. While international governing bodies such as the United Nations have established some, industry groups or multi-stakeholder efforts have independently developed others. These global guidelines are universally voluntary, functioning primarily to create efficiencies and comparability in the financial markets. These guidelines include reporting and disclosure guidance such as the Carbon Disclosure Project (CDP) and the Task Force on Climate-related Financial Disclosures (TCFD), as described below, as well as sets of aligning principles and commitments to provide common direction for solving environmental and social challenges (Table 14.1).

Government Managed Markets

Establishing regulated markets is one of the key tools that governments utilize to incentivize climate change conscientious and sustainable behavior more broadly. A regulated market is one in which government maintains some level of influence or control, such as by determining entrants into the market—for

Table 14.1 Global guidelines

Guideline	Description
UN Principles for Responsible Investment	In 2006, the United Nations launched the Principles for Responsible Investment (PRI). Since then, the PRI has become an independent organization working to understand the investment implications of ESG factors and to support its international network of investor signatories in incorporating these factors into their investment and ownership decisions
Network of Central Banks and Signatories for Greening the Financial System (NGFS)	The Network's purpose is to help strengthen the global effort to meet the goals of the Paris Agreement by enhancing the financial system to manage risks and mobilize capital for green and low-carbon investments. To this end, the Network defines and promotes best practices to be implemented within and outside of the Membership organizations and conducts or commissions analytical work on green finance[9]
Ceres Investor Network on Climate Risk	The Ceres Investor Network includes over 175 institutional investors, advancing investment practices, corporate engagement strategies, and key policy and regulatory solutions. Its initiatives include the Global Investor Coalition on Climate Change, Climate Action 100+, and The Investor Agenda[10]
We Mean Business	We Mean Business is a global nonprofit coalition working with businesses to take action on climate change. The participating organizations aim to "catalyze business leadership to drive policy ambition and accelerate the transition to a zero-carbon economy"[11]

Guideline	Description
RE 100	RE100 is a global corporate initiative of businesses committed to achieving 100% renewable electricity. Led by The Climate Group in partnership with CDP, RE100's purpose is to accelerate change toward zero-carbon grids at a global scale
Carbon Disclosure Project (CDP)	CDP organizes a global environmental disclosure system, supporting companies, cities, states, and regions to measure and manage their risks and opportunities on climate change, water security, and deforestation
Task Force on Climate-related Financial Disclosures (TCFD)	The Financial Standards Board's Task Force on Climate-related Financial Disclosures is a guideline for disclosure of climate-related financial risk by companies to investors. The TCFD Guidelines consider the physical, liability, and transition risks associated with climate change and what constitutes effective financial disclosures across industries
Sustainability Reporting Guidelines	These guidelines include: GRI's Sustainability Reporting Standards issued by Global Sustainability Standards Board (GSSB); The Organisation for Economic Co-operation and Development (OECD Guidelines for Multinational Enterprises); The United Nations Global Compact (the Communication on Progress); and The International Organization for Standardization (ISO 26000, International Standard for social responsibility)

example, requiring companies to trade emissions allowances. Alternatively, a regulated market can seek to influence the price of the tradeable commodity by throttling the number of emissions certificates entered each year. Regulated markets have become more commonplace, particularly in the United States following the success of this tool to address acid rain issues from utility air emissions in the 1980s and 1990s. Today there are several major regulated markets with extensive coverage, particularly in energy generation and climate change (Table 14.2).

Green Bond Guidelines and Standards

Green bonds refer to bond issuances where the proceeds of the bond will be used for sustainability-oriented projects. For example, a corporation or municipality might issue a bond in order to invest in energy efficiency measures. Green bonds have emerged as one of the most common financial instruments for promoting sustainability and addressing climate change. Several government and industry-led initiatives have sought to create standards for green bond issuance. The goal of these initiatives is to create comparability that inspires confidence that the sponsors issuing green bonds have demonstrable environmental benefit. They are also used to create efficiencies in the investment process to reduce the cost of searching for and screening green investment opportunities. China has issued a green bond regulation and the European Union has drafted a formal recommendation for regulating green bond issuance. In addition to the emerging government regulations, several voluntary standards for labeling green bonds have been adopted. The premise for voluntary labeling is to attract investors seeking demonstrable environmental benefit, either because they see potential benefits in green bonds, or because the investor has an internal policy mandate to purchase a certain level of environmental investments (Table 14.3).

Government Banking and Securities Regulation

In addition to the targeted regulations and initiatives above, there are a number of broader regulations for the banking and securities industries that have significant impacts, both negative and positive, on ESG and sustainability investments. Banking regulations exist at national and state/provincial levels of government, with jurisdiction over deposit-taking banks, principally to protect depositors against loss of their money, and in a larger sense to

Table 14.2 Government managed markets

Market	Description
Utility Regulation	Utility regulation varies and is specific to every country. In the United States, Investor-Owned Utilities are regulated by state-level public utility commissions, while government-owned utility agencies and departments are overseen by the government entity under which they exist (federal, state, and local). Europe does not have regulatory commissions but does regulate the utilities by law. Utilities around the world are generally departments of governments. Utility regulators can have direct and meaningful guidance on a utility's growth plan, system procurement plan, and acquisition of clean energy by ordering and overseeing auction processes, and approving the resulting Power Purchase Agreements (PPAs). Regulators also set auction terms, for example whether an auction is for renewable energy and not for coal, oil, or natural gas-fired power. Regulators also set forth required actions through rulemaking such as the Renewable Portfolio Standards (RPS) that require a certain percentage of the energy supply to come from renewables by a certain date
Public Utility Regulatory Policies Act (PURPA)	This U.S. law, enacted in 1978, established the practice of allowing non-utility organizations to develop, own, and operate renewable energy or efficient cogeneration power plants and sell the output electricity to utility companies and agencies, which is now common around the world

(continued)

Table 14.2 (continued)

Market	Description
Feed-In Tariff	This tool was the European version of the U.S. Public Utility Regulatory Policies Act of 1978 (PURPA), enacted originally by Germany at a national level in 1998, directing utilities to purchase the output electricity—"fed in" to the grid—from specified renewable energy projects. The mechanism spread across Europe and then to as many as 50 other countries around the world. By providing the equivalent of a government-sourced 20-year Power Purchase Agreements (PPA), the feed-in tariff was designed to attract low-cost long-term debt capital to Independent Power Producers—(IPP) and user-owned projects
Reverse Auctions	This approach has gained acceptance since about 2010 and has become the most common approach to awarding Power Purchase Agreements (PPAs) to renewable energy projects. It is called a "reverse" auction because the lowest-price bid wins
Carbon trading markets	Carbon trading, sometimes called emissions trading, is a popular market-based tool to limit emissions of greenhouse gases and other pollutants through a "market" mechanism. In carbon markets, firms trade emissions under cap-and-trade schemes or with credits that pay for or offset emissions reductions. The largest and most watched market is the European Trading System (ETS). Other such markets have been established in the United States (regionally), China (within provinces), and many other countries

Table 14.3 Green bond guidelines and standards

Regulation	Description
China Green Financial Bond Guidelines	China was the first country in the world to adopt official rules for the issuance of green bonds, with the People's Bank of China (PBoC) releasing its Green Financial Bond Guidelines at the end of 2015. The bank also developed an Endorsed Project Catalogue, providing detailed definitions of the types of projects that might be funded through green bonds, including energy and resource conservation, pollution reduction, clean transportation, clean energy, and ecological protection.[12] However, there is continuing controversy about this instrument, as qualifying bonds have been used to finance so-called clean coal
European Green Bond Standard	In 2019, the European Commission's special Technical Expert Group (TEG) published its Report on the EU Green Bond Standard. The TEG proposes that the European Commission establish a voluntary, non-legislative EU Green Bond Standard to enhance the effectiveness, transparency, comparability, and credibility of the green bond market and to encourage the market participants to issue and invest in EU green bonds
Green Bond Principles	Managed today by the International Capital Markets Association (ICMA), the Green Bond Principles (GBP) were originally promulgated in 2014 by a group of banks led by Citigroup, Bank of America Merrill Lynch, JP Morgan, and Credit Agricole. The Green Bond Principles are updated annually. These principles represent voluntary process guidelines recommending transparency and disclosure to promote integrity in the development and operation of the Green Bond market by clarifying the criteria for issuance of a green bond. The GBP were intended to, and continue to be, promulgated for broad use by the global bond market. The principles aim to: "provide issuers guidance on the key components involved in launching a credible Green Bond, aid investors by ensuring availability of information necessary to evaluate the environmental impact of their Green Bond investments and assist underwriters by moving the market towards standard disclosures which will facilitate transactions."[13]

(continued)

Table 14.3 (continued)

Regulation	Description
	The GPB genre has been expanded to include Social Bond Principles, Sustainable Bond Principles, and others. Importantly, these principles are bond market guidelines and are not regulated by governments, so they are being applied in a uniform manner globally
Climate Bond Standard	Hosted by the Climate Bond Initiative (CBI), a United Kingdom-based NGO, this certification scheme allows investors, governments, and other stakeholders to prioritize low-carbon investments. The standard relies on a scientific framework defining which projects and assets are consistent with a low-carbon and climate-resilient economy

ensure that banks are operating in line with best practices and actively stabilizing the financial system. Regulators oversee almost every element of banks' operations, but typically do not determine a bank's customers or the use of those customers' funds. Rather, since the 2015 Paris Agreement, governments and banking regulations have focused primarily on assessing the risks that banks are taking by lending to customers that in turn face climate risks. These assessment efforts are supported by related activities such as the TCFD, which calls for greater disclosure of climate risks faced by companies.

This field of securities regulation is so complex that a leading book, *Securities Regulation* by Louis Loss, Joel Seligman, and Troy A. Paredes costs $3815 (but shipping is free)![6] The principal focus of securities regulation is on publicly traded equity and debt capital markets (the "securities markets"). Within this market, securities regulators seek to ensure the disclosure of adequate information to enable investors to make informed investment decisions. Like banking regulators, securities regulators look at "how" things are done, but not so much "what" is done. That is, they specify and approve stock market offering documents, but not the business plan of the issuer. Likewise, securities regulators approve bond offering documents but refrain from opining on the likelihood of repayment—those opinions are given by the ratings firms such as Standard & Poors, Moody's, and Fitch, among others. On ESG matters, securities regulators are working on how to address climate risk and the attendant risk to each company by requiring disclosure of climate and sustainability-related risk. This process is tricky because, for example, an oil company that is deemed to have tremendous climate risk could choose tomorrow to sell its oil business and use the funds to acquire renewable energy companies. Unfortunately for ESG purposes, regulators must look backward for data, and not forward to possibilities.

Outlook on Sustainable Investing

Whereas sustainable investing and lending has been a niche product or department in otherwise broadly diversified financial firms, screening with ESG metrics has increasingly become a mainstream philosophy deployed across entire investment portfolios. However, funds are still managed according to long-standing, entrenched rules of risk/return, prudency, fiduciary duty, and other elements of financial industry culture that have supported the very industries that led us to global climate change! These entrenched rules are a barrier to mobilizing the financial sector's approximately $500 trillion of funds toward climate change solutions.[7]

Therefore, it is unlikely that sufficient capital will be deployed to stem climate change within our current financial system. A new system is needed. As Michael Bloomberg said as co-chair of the TCFD, "what gets measured gets managed."[8] The fundamental change required for the financial system is disclosure of *the use of funds* so investors can know how their funds will be used.

For example, in the issuance of a bond, the historical and current practice is to describe the use of funds from the bond as "general corporate purposes," with no disclosure of how the funds will be used. For government bonds, there are some general descriptions of the use of funds, but governments cannot be held to account because government budgets are set on an annual basis on the authority of legislatures. Under these practices, still in place today, there is no mechanism to determine if funds will actually be used to create environmental impact or benefit.

The world's success with clean power can serve as a model for addressing other global sustainability challenges. Policies were created by governments to create *policy-business models* that attracted project developers, investors, lenders, lawyers, engineers, accountants, and other actors. This goal was accomplished by creating the Independent Power Producer (IPP) Industry through public policies such as the Public Utilities Regulatory Policies Act (PURPA) in the United States and the Feed-In Tariff mechanism in Europe and around the world, that established the legal right of non-utility ("independent") entities to sell energy to the local (previously monopoly) utilities using Power Purchase Agreements approved by governments to create the assured revenue streams needed to allow project owners to make their debt payments. In addition, Renewable Portfolio Standards (RPS) in the United States and similar laws and goals around the world have created sustained market demand, supporting other policies such as tax credits and cash incentives to reduce the cost of clean energy to consumers, innovative financing structures such as "tax equity" to lower the cost of capital, and government guarantees to cover risks that could not be covered any other way—for example, in South Africa, Abu Dhabi, and other countries, the national governments have been required to sign the Power Purchase Agreements (PPAs), not just stand behind the signatures of the government-owned off-take utilities.

From a certain perspective, the direct purpose of public policy is to change the flow of money toward a designated purpose, whether the topic is security, education, housing, energy, or even climate change. While the focus is often predominantly on annual government budgets and spending, policy can be even more powerful in mobilizing the $500 trillion in the financial sector by establishing or changing the rules by which transactions are completed.

The opportunity remains for governments to use public policy to create business and investment models that create long-term revenues, and then reduce the risks for investors and lenders. These policy mechanisms must be created for each and every SDG and every climate change solution, one-by-one, as was done for clean energy. This endeavor will require tremendous collaboration and cooperation among the public, private, and financial sectors—involving governments, multilateral organizations, non-government organizations (NGOs), corporations, banks, investors, lawyers, scientists, engineers, and academia, among others.

Notes

1. Saxon, V. J. (2013, September 24). New York May Be Your Best Bet when Choosing the Governing Law and Forum for Your Cross-Border Contract. *Smarter Way to Cross.* Retrieved from https://www.lexology.com/library/det ail.aspx?g=e36cde01-e97b-46bc-869c-f595fcb42ea0#:~:text=New%20York%20may%20be%20your%20best%20bet%20when%20choosing%20the,for%20your%20cross%2Dborder%20contract; Rodriguez, M. M. (2020, January 28). New York v. Delaware Part 2: Which State Is Best for Governing Law? *Piliero Mazza Blog.* Retrieved from https://www.pilieromazza.com/blog-new-york-v-delaware-part-2-which-state-is-best-for-governing-law.
2. McCrone, A., Moslener, U., d'Estais, F., & Grüning, C. (2019). *Global Trends in Renewable Energy Investment 2019.* Frankfurt School—UNEP Collaborating Centre for Climate & Sustainable Energy Finance. https://wedocs.unep.org/bit stream/handle/20.500.11822/29752/GTR2019.pdf.
3. Torres-Rahman, Z., Baxter, G., Rivera, A., & Nelson, J. (2015). *Business and the United Nations; Working Together Towards the Sustainable Development Goals: A Framework for Action.* Published by Business Fights Poverty, The Harvard Kennedy School Corporate Social Responsibility School and the Sustainable Development Goals Fund. https://www.sdgfund.org/sites/default/files/business-and-un/SDGF_BFP_HKSCSRI_Business_and_SDGs-Web_Version.pdf.
4. Fink, L. (2020, January 14). A Fundamental Reshaping of Finance. *Blackrock.* New York.
5. Bloomenthal, A. (2020, April 17). Basel III. *Investopedia.* Retrieved from https://www.investopedia.com/terms/b/basell-iii.asp.
6. Wolters Kluwer. (2020). Securities Regulation. *Wolters Kluwer.* Retrieved from https://lrus.wolterskluwer.com/store/product/securities-regulation/.
7. Gadzinski, G., Schuller, M., & Vacchino, A. (2016). The Global Capital Stock. A Proxy for the Unobservable Global Market Portfolio. *SSRN Electronic Journal.* https://doi.org/10.2139/ssrn.2808438.

8. Drucker, P. (2017). *Practice of Management*. Routledge.
9. NGFS. (2019, September 13). Origin and Purpose. *NGFS*. Retrieved from https://www.ngfs.net/en/about-us/governance/origin-and-purpose.
10. Ceres. (2020). Ceres Investor Network on Climate Risk and Sustainability. *Ceres*. Retrieved from https://www.ceres.org/networks/ceres-investor-network.
11. We Mean Business Coalition. (2020). What We Do. *We Mean Business Coalition*. Retrieved from https://www.wemeanbusinesscoalition.org/about/.
12. Yu, K. (2016, January 29). Green Bonds, Green Boundaries: Building China's Green Financial System on a Solid Foundation. *IISD*. Retrieved from https://iisd.org/blog/green-bonds-green-boundaries.
13. International Capital Market Association. (2018). *Green Bond Principles: Voluntary Process Guidelines for Issuing Green Bonds*. Retrieved from https://www.icmagroup.org/assets/documents/Regulatory/Green-Bonds/Green-Bonds-Principles-June-2018-270520.pdf.

Index

© The Editor(s) (if applicable) and The Author(s), under exclusive
license to Springer Nature Switzerland AG 2020
D. C. Esty and T. Cort (eds.), *Values at Work*,
https://doi.org/10.1007/978-3-030-55613-6

9780305556129